P9-AFH-746

Clinical Specimens

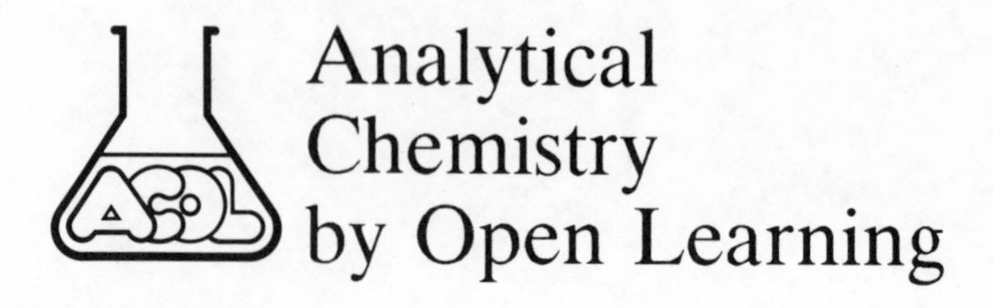
Analytical
Chemistry
by Open Learning

Titles in Series:

Samples and Standards
Sample Pretreatment
Classical Methods
Measurement, Statistics and Computation
Using Literature
Instrumentation
Chromatographic Separations
Gas Chromatography
High Performance Liquid Chromatography
Electrophoresis
Thin Layer Chromatography
Visible and Ultraviolet Spectroscopy
Fluorescence and Phosphorescence Spectroscopy
Infra Red Spectroscopy
Atomic Absorption and Emission Spectroscopy
Nuclear Magnetic Resonance Spectroscopy
X-ray Methods
Mass Spectrometry
Scanning Electron Microscopy and X-Ray Microanalysis
Principles of Electroanalytical Methods
Potentiometry and Ion Selective Electrodes
Polarography and Other Voltammetric Methods
Radiochemical Methods
Clinical Specimens
Diagnostic Enzymology
Quantitative Bioassay
Assessment and Control of Biochemical Methods
Thermal Methods
Microprocessor Applications

Clinical Specimens

Analytical Chemistry by Open Learning

Authors:
DAVID HAWCROFT and TERRY HECTOR
Both of
Leicester Polytechnic, UK

Editor:
ARTHUR M. JAMES

on behalf of ACOL

Published on behalf of ACOL, London
by
JOHN WILEY & SONS
Chichester · New York · Brisbane · Toronto · Singapore

Published by permission of the Controller of
Her Majesty's Stationery Office

Library of Congress Cataloging in Publication Data:

Hawcroft, David M.
Clinical specimens.
(Analytical chemistry by open learning)
Bibliography: p.
1. Diagnostic specimens—Collection and preservation.
I. Hector, Terry. II. Chapman, N. B. (Norman Bellamy), 1916– . III. ACOL (Firm : London, England)
IV. Title. V. Series. [DNLM: 1. Chemistry, Clinical—methods—programmed instruction. 2. Specimen Handling—programmed instruction. QY 18 H389c]
RB37.H34 1987 616.07'5 87-2151
ISBN 0 471 91396 0
ISBN 0 471 91397 9 (pbk.)

British Library Cataloguing in Publication Data:

Hawcroft, David
Clinical specimens.—(Analytical chemistry).
1. Chemistry Clinical
I. Title II. Hector, Terry III. ACOL
IV. Series

616.07'56 RB40

ISBN 0 471 91396 0
ISBN 0 471 91397 9 Pbk

Printed and bound in Great Britain

Analytical Chemistry

This series of texts is a result of an initiative by the Committee of Heads of Polytechnic Chemistry Departments in the United Kingdom. A project team based at Thames Polytechnic using funds available from the Manpower Services Commission 'Open Tech' Project have organised and managed the development of the material suitable for use by 'Distance Learners'. The contents of the various units have been identified, planned and written almost exclusively by groups of polytechnic staff, who are both expert in the subject area and are currently teaching in analytical chemistry.

The texts are for those interested in the basics of analytical chemistry and instrumental techniques who wish to study in a more flexible way than traditional institute attendance or to augment such attendance. A series of these units may be used by those undertaking courses leading to BTEC (levels IV and V), Royal Society of Chemistry (Certificates of Applied Chemistry) or other qualifications. The level is thus that of Senior Technician.

It is emphasised however that whilst the theoretical aspects of analytical chemistry can be studied in this way there is no substitute for the laboratory to learn the associated practical skills. In the U.K. there are nominated Polytechnics, Colleges and other Institutions who offer tutorial and practical support to achieve the practical objectives identified within each text. It is expected that many institutions worldwide will also provide such support.

The project will continue at Thames Polytechnic to support these 'Open Learning Texts', to continually refresh and update the material and to extend its coverage.

Further information about nominated support centres, the material or open learning techniques may be obtained from the project office at Thames Polytechnic, ACOL, Wellington St., Woolwich, London, SE18 6PF.

How to Use an Open Learning Text

Open learning texts are designed as a convenient and flexible way of studying for people who, for a variety of reasons cannot use conventional education courses. You will learn from this text the principles of one subject in Analytical Chemistry, but only by putting this knowledge into practice, under professional supervision, will you gain a full understanding of the analytical techniques described.

To achieve the full benefit from an open learning text you need to plan your place and time of study.

- Find the most suitable place to study where you can work without disturbance.

- If you have a tutor supervising your study discuss with him, or her, the date by which you should have completed this text.

- Some people study perfectly well in irregular bursts, however most students find that setting aside a certain number of hours each day is the most satisfactory method. It is for you to decide which pattern of study suits you best.

- If you decide to study for several hours at once, take short breaks of five or ten minutes every half hour or so. You will find that this method maintains a higher overall level of concentration.

Before you begin a detailed reading of the text, familiarise yourself with the general layout of the material. Have a look at the course contents list at the front of the book and flip through the pages to get a general impression of the way the subject is dealt with. You will find that there is space on the pages to make comments alongside the text as you study—your own notes for highlighting points that you feel are particularly important. Indicate in the margin the points you would like to discuss further with a tutor or fellow student. When you come to revise, these personal study notes will be very useful.

Π When you find a paragraph in the text marked with a symbol such as is shown here, this is where you get involved. At this point you are directed to do things: draw graphs, answer questions, perform calculations, etc. Do make an attempt at these activities. If necessary cover the succeeding response with a piece of paper until you are ready to read on. This is an opportunity for you to learn by participating in the subject and although the text continues by discussing your response, there is no better way to learn than by working things out for yourself.

We have introduced self assessment questions (SAQ) at appropriate places in the text. These SAQs provide for you a way of finding out if you understand what you have just been studying. There is space on the page for your answer and for any comments you want to add after reading the author's response. You will find the author's response to each SAQ at the end of the text. Compare what you have written with the response provided and read the discussion and advice.

At intervals in the text you will find a Summary and List of Objectives. The Summary will emphasise the important points covered by the material you have just read and the Objectives will give you a checklist of tasks you should then be able to achieve.

You can revise the Unit, perhaps for a formal examination, by re-reading the Summary and the Objectives, and by working through some of the SAQs. There are also fifty short Revision Questions at the end of the Unit which should quickly alert you to areas of the text that need further study.

At the end of the book you will find for reference lists of commonly used scientific symbols and values, units of measurement and also a periodic table.

Contents

Study Guide xiii

Bibliography xv

Acknowledgements xvii

1. Fluid Compartments of the Body 1
- 1.1. Extra- and Intra-cellular Fluids 1
- 1.2. Blood and Interstitial Fluid 6

2. Blood 8
- 2.1. Blood Components 8
- 2.2. Factors Affecting the Composition of Blood . . . 12

3. Collection of Blood Specimens 23
- 3.1. Types of Specimen 23
- 3.2. Actions of Anticoagulants 29
- 3.3. Patient and Specimen Identification 33
- 3.4. Changes in Blood Specimens During Storage . . 33
- 3.5. Hazards Associated with Blood Specimens . . . 43

4. General Points on Biopsy Samples 46
- 4.1. Organ Samples 47
- 4.2. Liquid Samples 51

5. Fetal Samples 68
- 5.1. Amniocentesis 68
- 5.2. Applications of the Fetal Biopsy Techniques . . . 76

Revision Questions and Responses 80

Self Assessment Questions and Responses 92

Units of Measurement 119

Study Guide

The human body is a biochemical machine of almost unimaginable complexity and largely to improve the efficiency with which its processes can be regulated, it is subdivided into compartments. The average non-scientist is aware that the body is subdivided into organs but this compartmentation continues with decreasing levels of magnification to tissues, cells, intracellular organelles and finally molecular complexes.

What is perhaps not quite so apparent is that there are different fluid compartments in the body also. The most obvious of these is the blood system with its roles of transporting food, wastes, gases and general metabolites around the body, but it should be borne in mind that these materials pass to and from the general body cells via an inter-cellular fluid which bathes the vast majority of body cells. Two other substantial fluid systems are those of the gut which at its beginning contains largely food materials and at its end waste products, and the urine which is also a mixture of wastes. Although these two fluids are strictly speaking outside the body they are nevertheless of great interest to us in medical diagnosis and disease monitoring.

One significant biological advantage of the organisation of the body into compartments is that it allows for a great deal of specialisation in function. It is obvious to the intelligent lay-person that nerve, muscle, skin etc are structurally different because they have different roles, and most important for us will have difference chemical compositions and processes. An important consequence of this specialisation is that if any but the most fundamental, and hence universal processes (eg cellular respiration) are to be studied, then we are frequently required to obtain specimens from those parts of the body that are specialised for that purpose. Thus it is that clinicians and others have devised ways of obtaining samples from a wide range of cells and tissues, including very specialised fluid compartments such as the synovial fluid of the joints, sweat, selected gut secretions, amniotic fluid around the fetus and even humoral fluid of the eye-ball. However it is very fortunate that many interesting metabolic changes result in measurable alterations in the chemical components of the blood and by investigating these changes we can commonly avoid having to obtain samples from specific organs, tissues or cells. It is for this reason that blood is the most frequently collected and investigated body material.

In this Unit we will discuss:

(*a*) the range of human specimens available, their roles, origins and specialised characteristics, and hence the reasons for our interest in them;

(*b*) the techniques used to obtain representative samples of these materials;

(*c*) the problems involved in sampling, transport, storage and analysing them.

Bibliography

General Texts

Bauer, J. D., *Clinical Laboratory Methods*, 9th ed Mosby, 1982.

Brown, S. Ed., *The Chemical Diagnosis of Disease*, Elsevier, 1979.

Hamilton, E., *Diagnostics* Nurse's Reference Library. Nursing 86 Books Springhouse Corp., 1986.

Lehninger, A. L., *Principles of Biochemistry*, Worth Publications Inc., 1982.

Montgomery, R. Ed., *Biochemistry – A Case-Orientated Approach*, 4th ed, C. V. Mosby Company, 1983.

Moran Campbell, E. J. Ed., *Clinical Physiology*, 5th ed, Blackwell Scientific Publications, 1984.

Tietz N., *Textbook of Clinical Biochemistry*, Saunders, 1986.

Varley H. Ed., *Practical Clinical Biochemistry*, 2 Vols, William Heinemann Medical Books, 1980.

Specialised References

Barson, A. J. and Davis J. A., *Laboratory Investigation of Fetal Disease*, Wright, 1981.

Borer W., *Chemical Analysis of Body Fluids other than Blood*, p115 in Chemical Diagnosis of Disease, Brown S. S., Mitchell F. and Young D., Elsevier, 1979.

Ferguson-Smith M. A., *Early Prenatal Diagnosis*, British Medical Bulletin, Vol 39, Churchill Livingstone, 1983.

Fraser C. and Watkinson L., *Patient, Specimens and Analysis as Potential Sources of Error*, p11 in Clinical Biochemistry Nearer the Patient, Marks V. and Alberti K., Churchill Livingstone, 1985.

Fraser C. G., *Interpretation of Clinical Laboratory Data*, Blackwell Scientific Publications, 1986.

McNeely M., *Urinalysis* in Gradwohl's Clinical Laboratory Methods and Diagnosis, ch22, Sonnewirth A., and Jarrett L. (ed), 8th ed Mosby, 1980.

Sandler M., *Amniotic Fluid and Its Clinical Significance*, Dekker, 1981.

WHO, *The Collection, Fractionation, Quality Control and Uses of Blood*, HMSO, 1981.

Acknowledgements

The data in Fig. 2.2a are taken from Varley H., Gowenlock A., Bell M., *Practical Chemical Biochemistry*, Vol. I, 5th ed., William Heinemann Medical Books, 1980.

Fig. 2.2b is based on a diagram from Brown S., *The Chemical Diagnosis of Disease*, Elsevier, 1979, permission applied for.

The data in Fig. 3.1a are taken from Erikssen K., Fox G., Trell E., Capillary–Venous Differences in Blood Glucose Values during the Oral Blood Glucose Tolerance test, *Clinical Chemistry* **29/5**, 973, 1983.

Fig. 3.1b is taken from a trade booklet produced for Sarstedt Ltd., Leicester, permission applied for.

Acknowledgments

1. Fluid Compartments of the Body

Overview

Movement continuously occurs between the fluids inside and outside the cells and abnormal concentrations of constituents in a clinical sample often reflect physiological and biochemical changes which have taken place in the body other than at the site from where the sample was collected. An understanding of the nature and composition of the different types of clinical samples and the way in which each relates to changes in specific organs, or the tissues generally, is helped by a knowledge of the main fluid compartments of the body and the factors which govern exchange of materials between them.

1.1. EXTRA- AND INTRA-CELLULAR FLUIDS

For a man weighing 70 kg roughly 28 dm^3 of fluid is intracellular and 14 dm^3 is extracellular. About 11 dm^3 of this extracellular fluid, known as the interstitial fluid, surrounds the cells of the organs of the body and is thus situated both anatomically and functionally between the intracellular fluid and the plasma in the vascular system. In addition there exists specialised compartments which contain fluids secreted by surrounding epithelial membranes. Cerebrospinal fluid which bathes the tissues of the central nervous system, the synovial fluid of the joints and secretions into the lumen of the gut are all examples of these transepithelial fluids. In health the volumes and compositions of the body's fluid compartments are maintained within narrow limits by the continuous internal exchange of water and solutes and by interaction with the outside environment via the kidneys, lungs, gut and the skin.

Osmotic pressure is the principal controlling influence on the movement of water across cell membranes with a tendency towards iso-osmotic conditions (equal osmotic pressure) between the intracellular and extracellular fluids.

SAQ 1.1a

To check your knowledge of osmotic pressure, and the features associated with it, write brief answers to the following questions.

(*i*) What is an ideal solution?

(*ii*) What does the osmotic pressure of a solution depend upon?

(*iii*) Osmotic pressure is one of the four colligative properties of a solution. Name the other three.

Substances such as urea and the respiratory gases are found at relatively equal concentrations inside and outside the cell since their distribution depends upon a simple thermodynamic equilibrium. The maintenance of very different intra- and extra-cellular solute concentrations requires some sort of active transport mechanism which involves the continued expenditure of energy by the cell.

The intracellular and extracellular concentrations of electrolytes relate to the cell membrane potential (positive on the outside relative to the inside) and may be predicted using the Nernst equation (Eq. 1.1a)

$$E = (RT/nF) \ln (c_e^+/c_i^+) \tag{1.1a}$$

Where E is the membrane potential in volts, n the valence of the ion R is the gas constant, T is the absolute temperature, F is the Faraday constant, ln is the natural logarithm and c_e^+ and c_i^+ the extra- and intra-cellular concentrations of the ion in mmol dm^{-3}.

At 37 °C the Nernst equation for a monovalent cation may be simplified;

$$E = 0.061 \log (c_e^+/c_i^+) \tag{1.1b}$$

It is necessary to reverse the extra and intracellular ion concentrations when dealing with an anion (Eq. 1.1c)

$$E = 0.061 \log (c_i^-/c_e^-) \tag{1.1c}$$

SAQ 1.1b

> The membrane potential of a red blood cell is 0.008 V and the plasma chloride concentration is 100 mmol dm^{-3}. Using the simplified version of the Nernst equation (1.1c), calculate the predicted intracellular chloride concentration for the red blood cell.

SAQ 1.1b

Since the cell membrane potential is balanced by the concentration gradient no electrochemical gradient exists for the chloride ion but this is not the case for many other ions. The extracellular concentration of sodium is about 145 mmol dm^{-3} which is far greater than its intracellular concentration (about 10 mmol dm^{-3} in skeletal muscle cells). This large electrochemical gradient is maintained by an active transport process known as the 'sodium pump' which may be inhibited by poisons or toxins and will deteriorate rapidly in any situations where the adenosine triphosphate (ATP) level in the cell is depleted.

SAQ 1.1c Can you name any situations which are liable to result in depletion of ATP levels within the cell?

The intracellular concentrations of potassium are approximately as predicted by the Nernst equation for some types of cells, but considerably different for others. In red blood cells the potassium concentration is far higher than expected due to a mechanism which may be linked with the sodium pump and results in active transport of potassium into the cell.

You should be aware that membranes are not static structures with pores of fixed dimensions. They can perform complicated and well-regulated functions which are often under hormonal control. Specific transport systems are found in membranes which translocate organic molecules such as glucose, regulate the movement of ions, and are responsible for the controlled release of metabolic products from the cell. For example glucose uptake by the cell is enhanced by insulin which has a counter-regulatory effect to the other carbohydrate-influencing hormones, ie glucagon, cortisol, epinephrine and the growth hormone.

At the cellular level insulin combines with a specific receptor on the cell membrane. The insulin-receptor combination triggers a 'second messenger' which stimulates the changes associated with insulin activity (Fig. 1.1a). The identity of this 'second messenger' is uncertain although indirect evidence indicates that it could be a small peptide.

Among the changes occurring in obesity there is a decrease in the number of insulin receptor sites in the cell membrane which reduces the sensitivity of the target cells to the hormone. Blood glucose and insulin concentrations are raised in obese persons, but usually return to normal levels when weight is lost and the number of receptor sites increases.

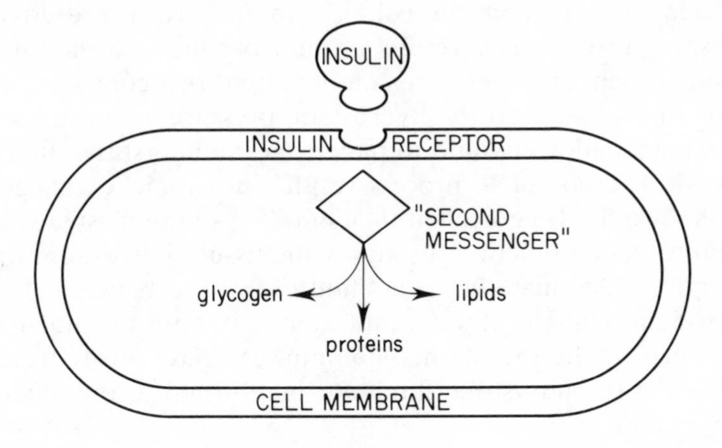

Fig. 1.1a. *Insulin activity at the cell membrane*

The human body is exceedingly complex and as a consequence has developed a large number of specialised systems for specific purposes. This specialisation exists at all levels of organisation from the molecular through the cellular to the individual organs of the body.

It is very relevant to the present discussion to realise that the fluid of the body is compartmented into specialised regions also. Though this occurs within cells (as the so-called 'organelles'), sampling of these for clinical purposes is still in its infancy. However the subdivision of extracellular fluids into compartments of specialised function is extremely important clinically since analysis of them can frequently provide much more specific information as to the organ and extent of disease states than might otherwise be the case.

1.2. BLOOD AND INTERSTITIAL FLUID

The interstitial fluid compartment consists of a network of collagen fibres supporting a gel which consists of hyaluronic acid and other complex polysaccharides. Quite considerable exchange occurs between the interstitial fluid and the blood plasma at both the arterial and venous ends of the vascular system. At the arterial end of the vascular system the hydrostatic pressure within the capillary lumen is sufficient to overcome the colloid osmotic pressure exerted by the plasma proteins. This results in fluid passing into the interstitial compartment and therefore coming into direct contact with the cells. At the venous end the hydrostatic pressure within the capillaries is lower, allowing fluid to pass from the interstitial fluid into the blood. The complete process results in a rapid exchange between the two fluids equivalent to about 75% of the plasma volume per minute, which functions to supply the tissue cells with nutrients and remove waste materials. The fluid exchanged between the two compartments consists of water and solutes, but not the blood cells. Protein, principally the plasma albumin, can pass slowly from the capillaries to the interstitial fluid and is returned to the blood via the lymphatics.

Summary

Fluid compartments of the body are identified as intracellular and extracellular, the latter being further subdivided into the interstitial fluid and the blood. Passage of water between these compartments is generally under osmotic control, while the distribution of solutes depends upon their nature, and the presence or absence of active membrane transport systems, which may be under the control of hormones.

A special relationship exists between the interstitial fluid and the blood. This allows for a considerable exchange of fluid between the two that ensures a ready supply of nutrients to the tissues and the removal of metabolic waste products.

Objectives

You should now be able to:

- identify the main fluid compartments of the body;
- identify osmotic pressure as the principal factor affecting the distribution of water between the fluid compartments;
- describe the various factors which influence the distribution of solutes between the fluid compartments;
- explain the special relationship which exists between the interstitial fluid and the blood.

2. Blood

Overview

The most commonly sampled body tissue is blood which consists of a variety of cells in a complex proteinaceous fluid called plasma.

In this section the composition of blood is described together with an account of the various factors, other than disease, that can influence the concentrations of its constituents.

2.1. BLOOD COMPONENTS

The blood volume of an adult is about 5 to 6 dm^3 of which almost half is composed of cells and the remainder is plasma. The red blood cells do not possess the organelles which could enable them to synthesise new proteins including enzymes. The absence of mitochondria leaves the red cells with relatively limited energy transformation processes. When a blood specimen is taken the red cells must then rely on a dwindling supply of glucose, and the presence of anticoagulants or preservatives may inhibit catabolic processes such as the glycolytic or the pentose phosphate pathways which are so vital to maintain normal membrane permeability upon which the survival of the cell depends.

In the body red cells are being continually replaced from the bone marrow as old cells disrupt and thus the blood contains a relatively constant mixture of cells in terms of age. From the time of collection of a blood sample there can be no cell replacement, but only disruption of the old and damaged cells with the release of their contents into the plasma. Compared with the red cells, the white blood cells (leucocytes) and platelets are present in small numbers, and should not be the cause of significant changes in the composition of a blood sample as a result of altered membrane permeability or their disruption.

About 90% of the plasma is water and the remaining 10% of the solute portion comprises about 70% protein, 20% relatively small organic compounds and 10% inorganic substances. Fig. 2.1a is a representation of the proportions of whole blood and plasma solute components.

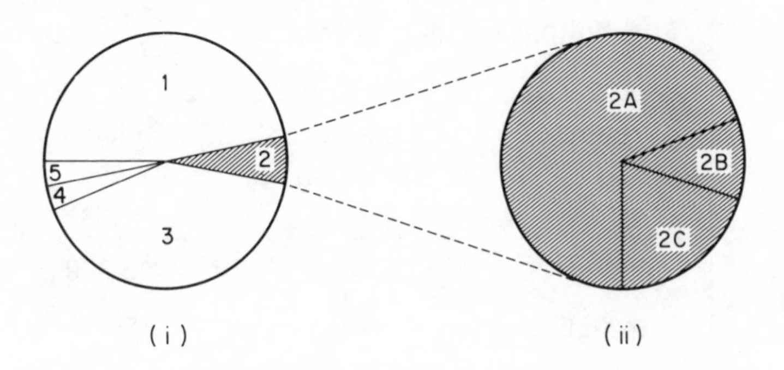

Fig. 2.1a. *Pie chart showing the percentage composition of whole blood and the solutes of plasma*

(*i*) Composition of whole blood

1. water
2. solutes
3. red blood cells
4. white blood cells
5. platelets

(*ii*) Composition of the solutes of plasma

2A. proteins
2B. inorganic solutes
2C. low molecular mass organic substances

The plasma proteins form a complex mixture with a variety of functions that include nutrition, control of the body water, buffer capacity, transport of other substances, blood coagulation, and immunological and enzymic activities. Some proteins found in the plasma have no particular role to perform there, having been shed from the tissues and only being in transit in the blood prior to their excretion or degradation by organs such as the kidneys or the liver.

SAQ 2.1a

Given the following information about the major protein fractions found in plasma, explain briefly which of them is likely to play the most important part in the control of blood volume.

Remember that the control of blood volume by the plasma proteins is due to their containment within the vascular fluid where they account for the colloid osmotic pressure.

Protein	Plasma levels ($g\ dm^{-3}$)	Relative molecular mass
albumin	40.0	66 300
α_1-globulins	4.5	40 000–60 000
α_2-globulins	6.5	100 000–400 000
β-globulins	8.5	110 000–120 000
γ-globulins	11.0	15 000–400 000
fibrinogen	3.0	340 000

SAQ 2.1a

The relatively small organic substances present in plasma consist of metabolites in the process of being transported between the various organs of the body. Water soluble substances such as urea or glucose are present in the free state in plasma while the non-water soluble substances such as lipids must be transported in combination with carrier proteins.

In health the concentrations of most of these substances remain reasonably steady except for the ones which are absorbed from the food (glucose, amino acids, fats, etc); these will therefore show transient increases in concentration following a meal. The intestinal absorption of fats gives the plasma a milky appearance due to the formation of chylomicrons which appear in the blood one to two hours after the meal was taken.

Inorganic substances such as sodium, potassium, calcium, chloride etc are usually present in the ionised state but there are exceptions. For example half the plasma calcium is in the ionised state (Ca^{2+}), about one tenth is complexed with substances such as citrate and bicarbonate and the rest is bound to plasma proteins. The subject of calcium binding with protein is discussed further in Section (2.2.6).

Under normal circumstances the greatest variation in the concentration of inorganic substances occurs as a result of respiratory exchanges in the lungs and other tissues. Most of the oxygen is taken up by haemoglobin within the red cells, but a great deal of the carbon dioxide released from the tissues is converted by carbonic anhydrase to bicarbonate which has a buffering effect. The oxygen and carbon dioxide/bicarbonate content of a specimen therefore very much depends on whether the blood is of venous, capillary or arterial origin and how the specimen was stored before the analyses were performed. The effects of storage on blood specimens are discussed in a later Section (3.4).

2.2. FACTORS AFFECTING THE COMPOSITION OF BLOOD

In addition to disease, the composition of blood is affected by factors that relate to individuals, their origins and habits and the environment in which they live. Some of these factors, for example age and sex, represent long-term influences on the concentrations of substances in the blood in contrast to factors responsible for short-term effects such as those observed after taking food. The long-term factors are of particular importance when selecting a sample population to use for the calculation of reference ranges for blood values. A knowledge of the influences of short-term factors may well affect decisions on practices such as the choice of blood specimen, the time when it should be collected and special precautions required in preparing the patient for the test.

2.2.1. Age and Sex

The physical changes associated with age are only too obvious, but these are also accompanied by changes in blood values. Four ages may be identified starting with the adaptation of a newborn infant to extra-uterine life, the childhood years leading to puberty, puberty to menopause for females or middle age for men and, lastly, old age. The start of each period, with the obvious exception of birth, cannot be accurately predicted by chronological age, but is rather a function of endocrine activity which varies from individual to individual.

Fig. 2.2a shows the effects of age and sex on the levels of some of the more commonly measured constituents of plasma. Differences in values observed between the sexes are generally greatest between puberty and menopause/middle age and are due to hormonal activities and the greater muscle mass of males over females. These differences tend to diminish with advancing old age for much the same reasons.

Constituent	Sex Difference	Effect of Increasing Age
Albumin	M>F	Falls.
Alkaline phosphatase	M>F (adults)	Marked rise in puberty adult levels when growth ceases. Rises during adult life; especially in menopause.
Alanine transaminase	M>F	Varies in different reports.
Aspartate transaminase	M>F	Falls to minimum at 30; later rise, especially F.
Bilirubin	M>F	Little change apart from post-menopausal rise.
Calcium	M>F	Falls, especially M.
Cholesterol	M>F	Rises, especially in post-menopausal F.
Creatine kinase	M>F	Falls in M; rises in F.
Creatinine	M>F	Slight rise.
Glucose	M>F	Rises.
γ-Glutamyltransferase	M>F	Falls in M; rises in F.
Hydroxybutyrate dehydrogenase	None	Rises in F.
Iron	M>F	Falls.
β-Lipoproteins	M>F	Little change.
Magnesium	M>F	Rises, especially F.
5′-Nucleotidase	None	Rises in F.
Phosphate	F>M	Little change apart from post-menopausal rise.
Potassium	None	Rises.
Sodium	None	Rises, especially F at menopause.
Thyroxine	F>M	Little change.
Total proteins	M>F	Little change.
Triacylglycerols	M>F	Rises.
Urate	M>F	Rises in F; steady or falls in M.
Urea	M>F	Rises.

Fig. 2.2a. *The effect of sex and age on various plasma constituents*

Biochemical changes during the menstrual cycle are not confined to the sex hormones. For example cholesterol is at its lowest in plasma at ovulation, increases immediately before menstruation and then continues to rise over the following week. Variations are also observed with many of the other commonly measured plasma constituents such as urate, creatinine, amino and fatty acids and the proteins (particularly fibrinogen).

Many changes are observed during the course of pregnancy. One profound influence on the blood levels of many substances is associated with the large increase in plasma volume which has a diluting effect, this also raises the rate of blood flow through the kidneys giving increases in the rate of excretion of substances such as urea and creatinine.

2.2.2. Ethnic, Dietary and Environmental Factors

Aside from the variations in frequency with which inherited defects occur, there are certain differences in blood constituents that are consistently found between the races. For example, negro children tend to have higher serum alkaline phosphatase activities at puberty than their caucasian contemporaries which reflects a higher rate of skeletal development at that stage of life.

In many instances it is not easy to separate the respective influences of race, diet and socioeconomic factors. It has been reported that Africans have higher serum urea levels than Europeans but, since urea production is directly related to protein intake, there is a strong possibility that diet, and not race, accounts for at least part of the difference. Reports have been published which attempt to establish correlations between levels of different plasma protein fractions and factors such as race, exposure to infection, dietary taboos and even income.

The levels of blood constituents are affected by both the calorific value of the food and its overall content. For example, the plasma levels of cholesterol and triacyglycerols of vegetarians are approximately two-thirds that of persons on mixed diets. Also, the particular content of a vegetarian diet tends to increase urinary pH thus affecting the excretion of metabolites such as urobilinogen and other pH dependant compounds including certain drugs.

Prolonged exposure to heat increases the volume of the extracellular fluid. In the blood this results in haemodilution since the increase in plasma volume is proportionally greater than that of the red cell mass. Heat also increases the amount of fluid lost as sweat and the electrolytes which it contains. Plasma sodium and chloride levels usually show little change as a result of sweating, but potassium concentrations may fall by as much as 10%. Extreme cold appears to be easier to adapt to than heat and result in fewer changes in blood composition. Changes in blood constituents have been associated with a range of other environmental factors. The hardness of the local water supplies affect serum cholesterol and triacyglycerol concentrations, and fluoridation of drinking water may reduce serum alkaline phosphatase activities. Exposure to environmental pollutants, such as lead and carbon monoxide from car exhausts, can lead to sustained increases of those substances in blood.

2.2.3. Circadian and Seasonal Variations

The concentrations of blood constituents vary according to factors such as sleeping patterns, rest periods, periods of physical activity, eating times, times of stress and even times of light and dark. The results of these factors are superimposed upon regular cyclical variations that are not present at birth but become established over a period of about three years. For example, the steroid hormone cortisol, which affects carbohydrate metabolism, is secreted in a cyclical pattern of about fifteen bursts in each twenty four hours. The duration and frequency of these bursts vary to provide plasma cortisol levels appropriate to the body's requirements at each particular time of the day. Fig. 2.2b shows that higher average plasma cortisol concentrations are found from just before mid-day and through the early afternoon period when people are generally more active. After long distance air travel involving changes of time zones, it may take several days for cortisol secretion to adjust to the new timing of daily activity; a similar situation is observed in shift workers in the period of the change-over between shift patterns.

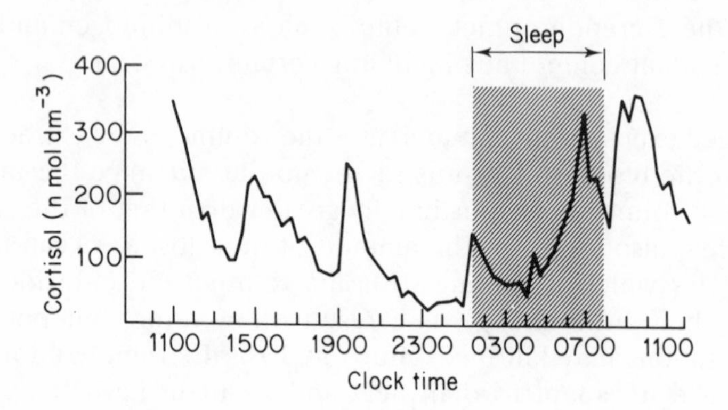

Fig. 2.2b. *Variation in the concentration of plasma cortisol throughout a day*

Seasonal changes seem to be principally associated with ambient temperatures and climatic conditions which in turn influence dietary factors such as the availability of fresh fruit and vegetables, and differences in the calorific value of the food intake. The extent and type of recreational activities and even the strength and duration of sunlight are also seasonally related. For example, the highest serum calcium values may be found in the summer as a result of greater ultra-violet irradiation of the skin increasing the conversion of 7-dehydrocholesterol to vitamin D; an important point since metabolites of the latter are responsible for increased mobilisation of calcium from the food.

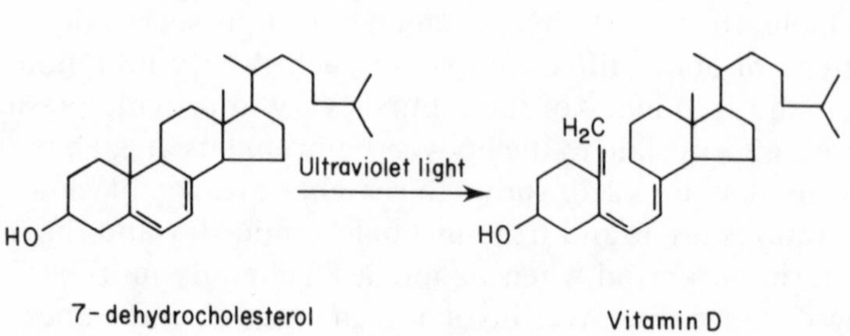

HO — Ultraviolet light → HO, H_2C

7-dehydrocholesterol — Vitamin D

Fig. 2.2c. *Structural formulae of vitamin D and its precursor*

SAQ 2.2a

Reports indicate that serum cholesterol concentrations tend to be higher in the winter than in the summer. Can you list three possible explanations for this finding? The formula given below for cholesterol may give you a clue for one of them.

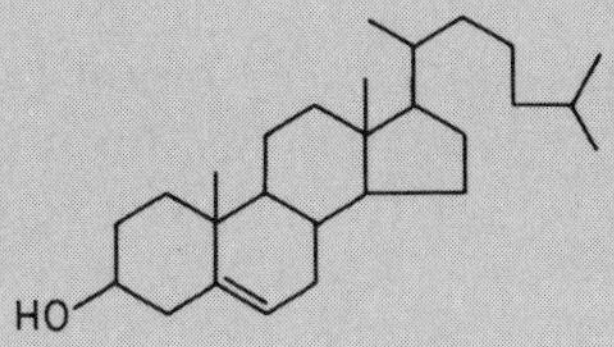

HO

Cholesterol

2.2.4. Stress and Physical Activity

Stress has a number of effects on the level of blood constituents. Cortisol increases in times of stress and plasma cortisol values have been measured as indicators of the relative degrees of stress that groups of people suffer under different conditions. Since cortisol is a glucocorticoid and thus affects carbohydrate metabolism, it comes as no surprise to learn that blood glucose levels also rise as a result of stress.

Admission to hospital and the general anxiety that a person feels about his or her state of health represent quite marked conditions of stress and contribute to the differences that are recognised between in-patients and the rest of the population. Acute anxiety may cause the patient to hyperventilate which has a direct effect upon substances such as lactate (increased blood levels) and causes acid/base imbalances.

Physical exercise can have a marked effect on the levels of many of the substances found in the blood including lactate, creatinine, fatty acids, glucose, amino acids, hormones and enzymes. Most of these substances return to normal levels shortly after stopping the activity, however enzymes, such as lactate dehydrogenase, released from the muscle cells may remain elevated for 24 hours or more.

Increased levels of urea, urate, creatinine and thyroxine have been found in the blood of athletes in training. These findings may be explained by the increased muscle mass built up by training and by increased cell protein turnover.

2.2.5. Body Size

The size of a person has little effect on most of the substances in blood, but it does determine the total amount of each of the substances in the body. In dynamic function tests it is usually necessary to take body size into account when calculating quantities of substances to be used either as stimulants, such as pentagastrin in studies of gastric function, or as challenge doses, such as glucose used in the glucose tolerance test.

Renal clearance, which is expressed as the volume of plasma that is completely cleared of a particular substance by the kidneys in unit time, is affected by the size of the patient. Differences in the body sizes of adults are generally not considered sufficient to affect clearance values significantly, but this is not the case with children. One simple correction for body size is to multiply the observed clearance figure by 1.73/A where 1.73 is the average normal body surface area for an adult in m^2 and A is the body surface area of the patient. The patient's body surface area is usually determined by the use of normograms based upon height and weight.

Obese individuals show differences in the levels of certain constituents of blood compared with lean persons. Lipids such as cholesterol and triacyglycerols are increased; the reduction of insulin receptors and its consequences on blood insulin and glucose values have been discussed (1.1).

2.2.6. Posture

Assuming an upright posture after a moderate period lying down results in haemoconcentration of the order of 10% to 20% over about 30 minutes. Some water and relatively small solutes of the plasma pass into the interstitial fluid leaving the cells and most of the proteins in the blood stream. Any substances that bind with proteins will also be affected and predicting just how their concentrations may change is not always easy as the following SAQ may demonstrate.

SAQ 2.2b

It was stated (2.1) that some of the plasma calcium is protein-bound and the rest is ionised or complexed with small substances such as carbonate or citrate.

A volunteer was asked to rest on a bed for three hours and a blood specimen was withdrawn at the end of that period while he remained in a recumbant position. Tests on the serum of that specimen showed a total protein concentration of 66.3 g dm^{-3} and a total calcium concentration of 2.30 mmol dm^{-3} of which 40% was protein-bound. The volunteer was asked to walk about for 20 minutes and a second blood specimen was then collected. ⟶

SAQ 2.2b (cont.)

Make an estimate of the total serum calcium concentration of the second specimen which had a total serum protein value of 76.5 g dm^{-3}.

Note any assumptions that are necessary when making your estimate of the new calcium concentration.

Haemoconcentration is not the only result of changes of posture. The stomach empties more slowly in recumbent persons which affects the rate at which substances such as glucose can be absorbed into the blood from the food or during function tests. Renal function is also affected by posture.

2.2.7. Recent Food

In addition to circadian variations quite significant changes in concentration of blood constituents occur after meals. The extent of the changes depends on the particular test substance, the content and size of the meal and the time at which the specimen was collected after the food was taken.

Blood glucose concentrations are generally increased for some 2 hours after a meal but should return to fasting levels in 3 hours. If the meal had a high protein content increases in plasma urea, urate and phosphorus can be found even 12 hours later. Fat absorption leads to the presence of chylomicrons in the blood which reach peak concentrations some 2–4 hours after the meal and give the plasma or serum a cloudy appearance that can cause problems when spectrophotometric assay methods are used.

The actual process of digestion is also responsible for changes in the blood. For example, secretion of gastric acid causes the so-called 'alkaline tide' in the venous blood leaving the stomach. This gives a mild alkalosis, but it is still sufficient to reduce the ionised plasma calcium concentration by as much as 0.05 mmol dm^{-3} one hour after the meal.

Fasting blood specimens are usually requested for analyses of substances such as glucose or calcium, which show unpredictable changes in concentration after food has been taken.

2.2.8. Drug Administration

There is a strong possibility that a drug will affect tissues, other than the one which is affected by disease, and that the metabolic consequences of such actions will lead to changes in the blood. Exactly what those changes will be depends upon the drug, its bioavailability, the route of administration, dosage, and whether it is used alone or in combination with other drugs.

It is important to consider both the pharmocological and the methodological effects of interference by a drug on biochemical assays. A laboratory might well ask physicians to enter a complete list of the drugs taken by a patient on the request form in order to be able to make allowances for these.

In addition to the use of drugs for therapeutic purposes, the intake of beverages containing caffeine such as coffee, alcohol consumption, smoking and other drugs of addiction also affect metabolic processes and thus blood values.

Summary

Blood contains red cells, white cells and platelets suspended in plasma which is a mixture of proteins, smaller organic and inorganic molecules and ions. Many factors, other than disease, affect the concentration of blood constituents. Some have long term effects such as age, sex, the habitual diet, race and the environment, while others, such as body size, posture, meals, physical activity and conditioning, stress, drugs and specific foodstuffs, have more transient effects on blood values.

These factors should be taken into account when designing sampling programmes for the calculation of reference ranges and when making decisions on the most suitable sample to collect, its time of collection, and any special arrangements regarding preparation of the patient for the test. The physician may also wish to take note of such factors in his interpretation of the test results.

Objectives

You should now be able to:

- describe the composition of blood;

- discuss how the composition is influenced by:

  age and sex,
  race, diet and the environment,
  circadian and seasonal variations,
  stress and physical activity,
  body size,
  posture,
  recent intake of food,
  drugs, stimulants etc;

- recognise the importance of these *in vivo* factors with respect to reference ranges and sampling.

3. Collection of Blood Specimens

Overview

Arterial, venous and capillary blood specimens may be collected from the patient and subsequent analyses performed on either whole blood, plasma or serum. This section discusses the choices which are available, their respective advantages and disadvantages and the factors which affect the composition of specimens from the time of their collection to when they are analysed.

3.1. TYPES OF SPECIMEN

3.1.1. Site of Collection

Biochemical assays can be performed on arterial, capillary or venous blood although arterial blood specimens are taken only in special circumstances, since the collection procedure carries a higher degree of risk to the patient than does venepuncture or capillary sampling by skin prick. Venepuncture is a procedure that also carries a small element of risk to the patient, and must only be performed by properly trained staff.

In considering the function of the arterial supply as the bringer of oxygen and nutrients to the tissues, and the venous system as the remover of carbon dioxide and other metabolic waste, it is axiomatic that the concentrations of the constituents of arterial and venous blood should differ in some respects. As long as the collection site is warmed and stasis is avoided, capillary blood closely approximates to arterial blood and can be used in its place for biochemical investigations of, for example, blood gases and acid/base status. Blood from a heel prick is the usual specimen from infants and finger tip or ear lobe capillary samples are commonly used when the collection of venous blood from adults is difficult or inconvenient.

Collection of capillary blood is relatively simple, particularly with the use of commercially available skin puncture devices; the technique is easily acquired by most people with the minimum of instruction. Aside from only being able to collect small sample volumes, there is a tendency for capillary blood specimens to suffer haemolysis more readily than does venous blood, and for contamination with tissue debris to occur.

Differences in the concentrations of the constituents of capillary and venous blood are minimal with the exception of blood gases and their products and glucose. In a fasting person the differences between capillary and venous blood glucose levels are small but this is not so after meals.

Eriksson and his colleagues performed glucose tolerance tests by giving volunteers standardised oral doses of glucose and then collecting venous and capillary blood specimens at intervals of up to two hours. The mean results which they obtained (Fig. 3.1a) show that under fasting conditions blood glucose values are about the same for venous and capillary samples, but that after the glucose was given there are significant differences (statistically confirmed at $p < 0.0005$ by the Student's t-test).

Blood specimen	Time after ingestion (minutes)					
	0	20	40	60	90	120
	Blood glucose concentrations (mmol dm^{-3})					
Capillary	5.06	7.58	10.02	10.15	8.67	7.08
Venous	4.98	6.38	8.56	8.64	7.49	6.21
Difference	0.08	1.02	1.46	1.51	1.18	0.87

Fig. 3.1a. *Mean glucose concentrations in capillary and venous blood specimens (ref: Clinical Chemistry,* **29** *(1983) 993)*

SAQ 3.1a

Based on the results of Eriksson and his colleagues (Fig. 3.1a) suggest a set of recommendations relating to the collection of capillary and venous blood specimens when performing glucose tolerance tests.

Venous blood is usually the specimen of choice. It is most commonly collected from the median cubital vein in the antecubital fossa (crook of the elbow) using either a sterile syringe and needle or a needle attached to a special vacuum tube. Fig. 3.1b is a diagram of the 'Monovette' tube currently marketed by the Sarstedt Company and shown in use as a vacuum tube (this product may also be used as a syringe). The vacuum tube also serves as a specimen container whereas blood collected into a syringe must be transferred to a separate bottle or tube. Blood should be added slowly to the specimen container, and mixing needs to be gentle if haemolysis is to be avoided.

A tourniquet is commonly used to distend and aid location of the veins, but it can result in changes in venous blood values and such changes tend to increase with the time of application. Before the needle is inserted into a vein the skin is swabbed with an alcohol-containing fluid and then allowed to dry. This does not normally cause any problems but there are always exceptions to every rule. If blood ethanol assays are to be performed the use of ethanol in the swab fluid may lead to falsely high values whereas the alternative, propan-2-ol, may give low values in gas liquid chromatographic assays where it is employed as the internal standard.

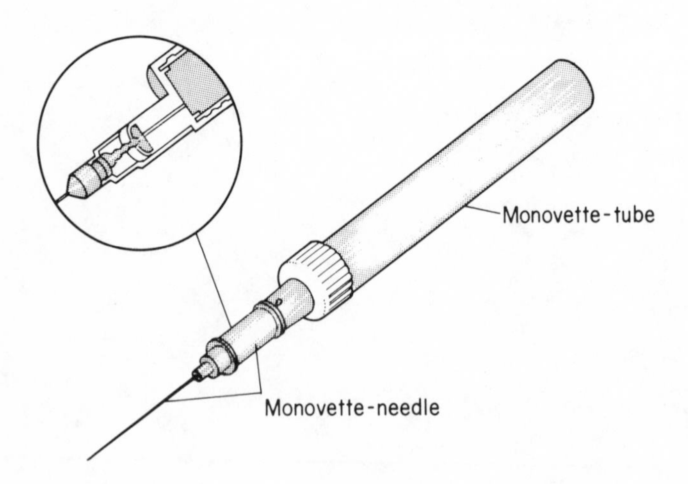

Fig. 3.1b. *The Sarstedt Company variable collection unit showing its use in the vacuum tube mode*

It is preferable for the patient to be comfortably seated for 20 minutes before the specimen is collected, or lying down if sitting is not possible, in order to standardise the haemoconcentration effects of posture. Blood should not be taken from an arm into which fluids, eg transfusion blood, saline etc are being infused, and this should certainly not be done without the consent of the physician. If a woman has had a mastectomy, lymphostasis may affect the composition of blood in the arm veins on that side of the body and thus specimens are best collected from the other side.

SAQ 3.1b

A tourniquet is applied to a volunteer's arm and a needle inserted in the median cubital vein. Exactly 5 cm^3 of blood is allowed to run at about 10 cm^3 min^{-1} into each of four bottles labelled 1, 2, 3 and 4 respectively. The blood samples are allowed to clot and albumin determined in the serum. Results as follows:

Specimen number	1	2	3	4
Albumin (g dm^{-3})	43	44	46	49

(*i*) Can you explain the increase of serum albumin in terms of the tourniquet effect?

(*ii*) Which of the samples is likely to be the most representative of the circulating blood?

(*iii*) In drawing up recommendations for venous blood collection are there any points which you would make relating to volumes of blood to be collected for specific analyses?

Remember that the tourniquet will increase intravascular pressure, and this in turn will increase filtration pressure across the capillary walls.

SAQ 3.1b

3.1.2. Whole Blood, Plasma or Serum

Whole blood is the preferred specimen for some analyses, eg blood gases, ammonia and lead, and can be used for substances which are more or less evenly distributed between cells and plasma such as glucose, urea or lactate. There are two distinct advantages of using whole blood in that it provides a greater volume of sample than can plasma or serum, and no separation stage is required which gives economy of both time and equipment.

Unfortunately the contents of blood cells tend to have interfering effects on many of the analytical methods used in biochemistry, eg spectroscopic methods are particularly prone to interference from haemoglobin.

The plasma or serum levels of many substances are used to diagnose and monitor disease, eg lactate dehydrogenase activities for studies of liver disease and myocardial infarction. These same substances are also present in large amounts in the blood cells which, if included as part of the analysed sample, would vastly reduce the sensitivity of the test to changes in plasma levels. Failure to centrifuge adequately anticoagulated blood may leave platelets suspended in the plasma. Such plasma can give falsely high values for some constituents such as lactate dehydrogenase, depending on the assay method.

The use of serum not plasma as the sample of choice by many laboratories is determined by several factors. Some investigations are interfered with, or confused by, plasma constituents, eg protein fractionation by the fibrinogen in plasma. Anticoagulants may influence the assay procedure either by affecting the form in which the test substance is present or by interfering directly with the test reaction. Finally, small fibrin clots often form in stored plasma and may block the narrow bore tubing used on many automated analytical instruments.

It is possible to separate plasma from the cells immediately after the blood is collected, compared with a minimum time of 15 to 30 minutes that needs to be allowed for clot formation and contraction, before serum can be harvested. Larger volumes of plasma can be separated from blood than from serum although much of this increased volume is due to the space occupied by fibrinogen. In some cases plasma is preferable to serum to avoid artefacts of the clotting process. For example, potassium release from the cells during clotting gives serum levels some 0.2 mmol dm^{-3} above those of plasma in healthy individuals, but in patients with abnormal blood cell populations this figure can increase to 1.6 mmol dm^{-3} higher.

3.2. ACTIONS OF ANTICOAGULANTS

Blood coagulation is the result of a complex sequence of enzymic reactions that terminate with the cleavage by thrombin of two small peptides from fibrinogen to yield the insoluble polymer, fibrin. The fibrin strands are then cross-linked by a further thrombin-initiated reaction. Calcium ions are required for the activation of several of the reactions in this sequence.

Anticoagulants lower the effective concentrations of one or more of those factors which are actively involved in the clotting process and are very commonly used in blood sampling to provide samples which are suitable for analysis. The type and quantity of anticoagulant used should be carefully considered with respect to the characteristics of the test substance and to the intended method of analysis. Some of the anticoagulants in common use by clinical laboratories are described below.

3.2.1. Heparin

Heparin is a naturally occurring anticoagulant which consists of a highly sulphated glycosaminoglycan (Fig. 3.2a), with a relative molecular mass of about 10 000 and is available as sodium, potassium, ammonium, calcium and lithium salts.

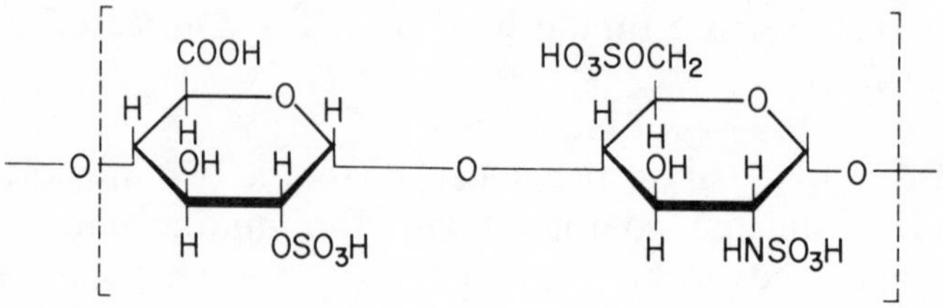

Fig. 3.2a. *The structural formula of the repeating unit of heparin*

Heparin is believed to promote the combination of a naturally occurring plasma protein called antithrombin III with various blood coagulation factors yielding inactive complexes and thus preventing the conversion of prothrombin to thrombin.

It is used at about 2 mg cm^{-3} of blood to be collected. In clinical chemistry the potassium and sodium salts should be avoided for plasma electrolyte studies (plasma sodium concentrations may be about 2 mmol dm^{-3} too high). Insignificant differences in electrolyte results are obtained when lithium heparin is used as the anticoagulant for flame photometric methods with a lithium external standard, but the same is said not to be true in the case of methods using direct reading ion-selective electrode systems.

Other analyses which are said to be affected by heparin are those for calcium which involve EDTA complexing methods, assays for enzymes such as hydroxybutyrate dehydrogenase and acid phosphatase, and for hormones such as thyroxine where binding to carrier globulins is reduced.

3.2.2. Ethylenediaminetetracetic Acid (EDTA)

EDTA is a chelating agent which binds calcium and leaves insufficient in the ionised form to take part in the clotting process. Used as the disodium or dipotassium salt at final blood concentrations of 1–2 mg cm^{-3}, it is the anticoagulant of choice for haematological investigations since cell morphology is well preserved. In clinical chemistry laboratories it is rarely used although its effect on tests other than those involving metals or the presence of metallic cofactors seems minimal.

3.2.3. Sodium Citrate

Sodium citrate is used as one volume of a 3.4 or 3.8 g dm^{-3} solution to nine volumes of blood thus introducing an immediate 9/10 dilution of all blood constituents. Sodium citrate gives good preservation of the 'factors' (enzymes, fibrinogen etc) which are involved in the blood coagulation process. It is often the anticoagulent of choice for investigations of blood clotting diseases but is rarely used otherwise.

3.2.4. Oxalates

Oxalates act by forming insoluble calcium oxalate and thus effectively removing calcium ions from solution. Various salts are available of which potassium oxalate is the most widely used at concentrations from 1 to 2 mg cm^{-3}. Whenever anticoagulants are used a sufficient quantity of blood should be added to the specimen container to ensure that the final concentration of anticoagulant is correct, but this is particularly important with potassium oxalate since haemolysis is likely to occur at concentrations exceeding 3 mg cm^{-3}.

Potassium oxalate causes cell shrinkage with dilution of the plasma and exchange of electrolytes, and is therefore unsuitable for investigations of these. However it has a tendency to inhibit many enzymes of clinical interest including acid and alkaline phosphatases, amylase and lactate dehydrogenase.

SAQ 3.2a

The following set of plasma values were obtained for an apparently normal person. There are grounds for suspecting that the blood specimen was collected into an inappropriate anticoagulant.

Try to identify which anticoagulant(s) described previously could have been used.

Analyte	Units	Result	Reference value
Potassium	mmol dm^{-3}	6.3	3.7–5.2
Sodium	"	139	135–146
Calcium	"	1.9	2.3–2.7
Bicarbonate	"	24	22–26
Chloride	"	103	98–108
Phosphate	"	1.12	0.7–1.4
Urea	"	4.9	3.3–7.5
Total protein	g dm^{-3}	72	68–79
Albumin	"	44	40–48

3.3. PATIENT AND SPECIMEN IDENTIFICATION

It is common practice for each in-patient (hospitalised patient) to wear an identification bracelet giving his or her name and a designated hospital number. Positive patient identification can be achieved by asking for his or her name and cross checking the answer (assuming the patient is conscious and lucid) with information given on the bracelet and on the bedside notes. If any doubt exists, the phlebotomist (blood-taker) should ask the nurse who has responsibility for the patient to confirm the identification.

Having verified the patient's identity and that he or she has been properly prepared for the test (eg he or she is fasting should fasting be necessary), the sample can be collected. The sample container should be clearly labelled at the bedside with the patient's name, hospital number, the date and time of collection, and the ward or room number. It should then be dispatched together with the sample request form to the laboratory without further delay.

3.4. CHANGES IN BLOOD SPECIMENS DURING STORAGE

It is rare for specimens to be analysed immediately after they have been collected and yet even a short delay can result in changes in the concentration of some constituents.

Changes which occur in blood specimens during storage can be broadly classified as follows:

- primary or secondary effects associated with gaseous exchanges;
- changes resulting from the continuation of metabolic reactions within the blood cells;
- haemolysis or membrane permeability changes;
- changes due to denaturation and loss of biological activity.

3.4.1. Changes Associated with Gaseous Exchange

Contact of a blood specimen with the air brings about changes in pO_2, pCO_2 and related molecules and ions. The type and extent of these changes are dependent upon the temperature and time of storage and the amount of contact with the air. Shaking during sampling and transport substantially increases this contact with air.

SAQ 3.4a

5 cm^3 of venous blood is placed in an anticoagulant-containing bottle which has a total volume of 15 cm^3. This leaves a considerable air space above the specimen.

Try to predict whether the concentrations of the following substances are likely to increase, decrease or remain constant in both the red cells and the plasma as a result of mixing the specimen with air. The behaviour of chloride ions is not easily identified. Try to remember that it is not just the balance of each individual species between cells and plasma which is important, but also the balance between charged species.

Substance	Cells	Plasma
oxygen (O_2)		
carbon dioxide (CO_2)		
bicarbonate (HCO_3^-)		
chloride (Cl^-)		

The following information may help your decisions: →

SAQ 3.4a (cont.)

(*i*) the red cell membrane is permeable to CO_2, O_2, HCO_3^-, Cl^-, and H^+;

(*ii*) loss of a substance from cells will be at least partially compensated by movement of that substance from the plasma and vice versa;

(*iii*) although intra- and extra-cellular concentrations of individual substances may vary, the balance between cations and anions on either side of a cell membrane is usually maintained;

(*iv*) red cells contain the enzyme carbonic anhydrase which catalyses the reversible reaction between hydrogen and bicarbonate ions to form carbonic acid leading to carbon dioxide and water:

$$H^+ + HCO_3^- \rightleftharpoons H_2CO_3 \rightleftharpoons CO_2 + H_2O$$

Remember that enzymes increase the rate of reaction towards equilibrium states;

(*v*) when oxygen combines with haemoglobin a proton (H^+) is displaced. The reverse occurs at low pO_2 values when a proton is taken up by haemoglobin as oxygen is released.

(*vi*) compared with room air venous blood is relatively deficient in oxygen and rich in carbon dioxide.

SAQ 3.4a

The gaseous exchange that occurs when venous blood is exposed to air causes an increase in pH. Total concentrations of constituents are not likely to be affected by the pH change, but the form in which they exist may be. For example, the increase in pH is accompanied by a lowering of the concentration of ionised calcium and a corresponding increase in protein binding.

3.4.2. Continued Cellular Activity

When blood is removed from the body the rate of cellular activity may slow down but it by no means stops. Since red cells greatly outnumber other blood cells they tend to be the major source of changes. However significant effects are seen in the blood of patients with diseases in which white blood cells proliferate, eg leukaemia.

Cellular activity requires energy in the form of compounds such as ATP which can be generated by the oxidation of glucose. In the body the blood glucose concentration is maintained from the glycogen reserves of the liver and by intestinal absorption of the sugar. These sources are denied to the isolated blood specimen, and it is therefore not surprising that glucose levels show a steady decline (about 7% per hour at room temperature).

SAQ 3.4b

Respiration is often summarised as the oxidation of glucose to carbon dioxide and water:

$$C_6H_{12}O_6 + 6\,O_2 \rightarrow 6\,CO_2 + 6\,H_2O + \text{energy}$$

In stored blood however, the main product of glucose catabolism is lactic acid:

$$C_6H_{12}O_6 \rightarrow 2\,C_3H_6O_3$$

Which of the following statements offers the best explanation for the difference between glucose breakdown in the blood compared with that in other tissues?

(*i*) Red cells metabolise glucose by a different pathway from that used by other cells of the body.

(*ii*) Glucose catabolism is a two stage process. The first stage takes place in the blood, and the second stage in the tissues where the lactic acid is broken down to carbon dioxide and water.

(*iii*) Red cells lack mitochondria, hence they are unable to oxidise glucose completely.

(*iv*) Complete oxidation of glucose to carbon dioxide and water requires the presence of oxygen. In stored blood the supply of oxygen is limited since the cells settle at the bottom of the bottle and are separated from the air by a layer of plasma.

SAQ 3.4b

3.4.3. Haemolysis

In vitro haemolysis (red cell destruction) which is only just detectable to the naked eye accounts for an approximate 0.2% increase in the plasma volume. The volume increase must have a diluting effect on all plasma constituents, but it is of little significance even for those substances which are only present extracellularly.

Haemolysis causes major problems with plasma assays of substances which are present in very large amounts in the cells. Based on the values given in Fig. 3.4a, a plasma which just shows signs of haemolysis would have a lactate dehydrogenase activity some 30% higher than that of a non-haemolysed specimen.

Constituent	Units	Red Cells	Plasma
Potassium	mmol dm^{-3}	100	4
Sodium	"	15	144
Calcium	"	0.25	2.5
Chloride	"	50	102
Bicarbonate	"	10	25
Phosphate	"	4.0	1.3
Urea	"	4.0	5.6
Glucose	"	4.2	5.2
Lactate dehydrogenase	U dm^{-3}	30000	180
Aspartate amino-transferase	"	500	30

Fig. 3.4a. *Concentrations of some constituents of red blood cells and plasma*

(U is the International Unit of enzyme activity)

Haemoglobin, by virtue of its inherent colour, can be troublesome in spectrophotometric assays in the visible regions of the spectrum. One assay that is particularly prone to interference by haemoglobin is serum or plasma bilirubin determination on blood from new born infants. The assays are commonly performed by absorption spectrophotometry on serum or plasma diluted in phosphate buffer solution. Blood is usually obtained by heel prick and such specimens often show haemolysis. The problem caused by the presence of haemoglobin is the topic of the following SAQ.

SAQ 3.4c

Absorption spectra for bilirubin and haemoglobin in phosphate buffer solutions are shown in Fig. 3.4b and 3.4c respectively.

Can you suggest:

(*i*) suitable wavelength for reading the absorbance value of bilirubin, and ⟶

SAQ 3.4c (cont.)

(*ii*) a simple method of correcting the bilirubin absorbance for interference by any haemoglobin which might be present in a serum sample.

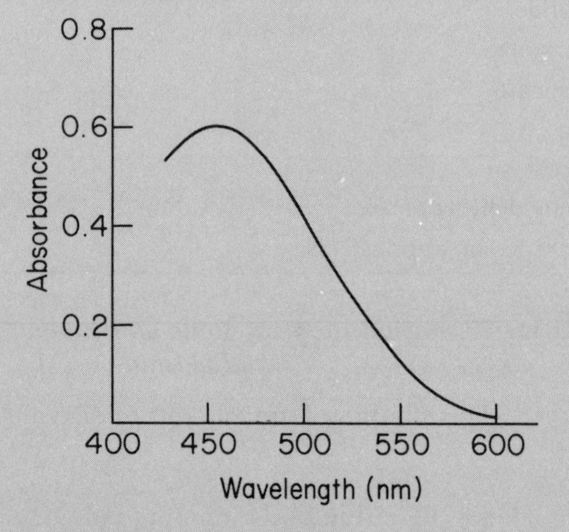

Fig. 3.4b. *Absorption spectrum for bilirubin*

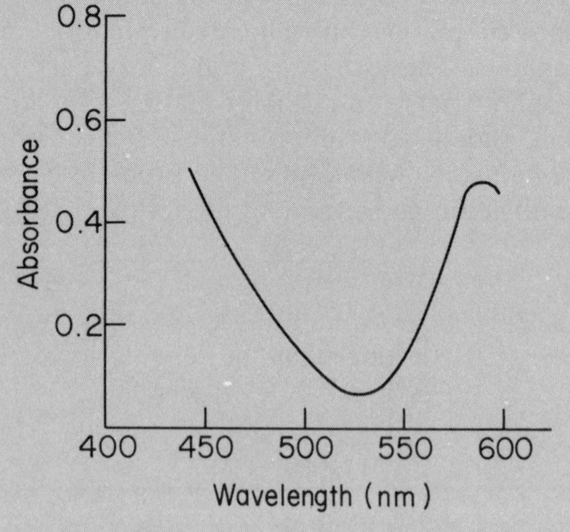

Fig. 3.4c. *Absorption spectrum for haemoglobin*

SAQ 3.4c

Haemolysis is an indication that red cells have broken up, but it is possible for cell contents to leak out from intact cells during storage. At 4 °C the active transport mechanism which maintains a high intracellular potassium concentration is depressed and there is leakage of the ion into the plasma. It is therefore better to store blood intended for plasma-potassium determinations at room temperature rather than in the refrigerator. Room temperature however may not be suitable for other constituents and a far better practice is to keep the storage time as short as possible before the plasma is taken off.

3.4.4. Denaturation and Loss of Biological Activity

Denaturation or the loss of biological activity of constituents can be minimized and the quality of results improved if specimens are transported to the laboratory without delay and the plasma or serum separated off as soon as possible.

Many constituents of serum or plasma remain stable for several days at 4 °C and for months when stored in a deep-frozen state. Some constituents are not so robust, and demand that special attention is paid to the collection, transport, preservation and storage conditions.

One of the most labile compounds in plasma is adrenocorticotrophic hormone (ACTH, corticotrophin). It is adsorbed onto glass surfaces, is very unstable in whole blood and is destroyed by proteolytic plasma enzymes that are activated by the freezing and thawing of samples. A lightly heparinized plastic syringe is best used to collect the blood specimen which is chilled by placing in iced water and then taken immediately to the laboratory. It is centrifuged for 15 minutes at 1000–1500 g at 4 °C and the plasma transferred to a second plastic tube. The plasma is recentrifuged for 10 minutes at 6000 g at 4 °C and the supernatant stored in a deep-frozen state until the analysis can be performed. The second centrifugation removes formed structures which, if frozen and thawed, may rupture and release proteolytic enzymes that denature ACTH.

Not only do different enzymes vary in their stability in stored blood, plasma and serum but variability is also seen between isoenzymes (physico-chemical variants of the same enzyme). Acid phosphatases are generally not stable enzymes. The prostatic isoenzyme of acid phosphatase is particularly labile in serum and may lose over half its activity in one hour at room temperature. This lability seems to be related to temperature and pH. Stability is improved by storing in the refrigerator after acidification with disodium citrate monohydrate at a final concentration of 10 mg cm^{-3}.

The five isoenzymes of lactate dehydrogenase (LDH) vary in their sensitivity to cold, LD-4 and LD-5 are the most labile of the five and can show complete loss of activity overnight in tissue extracts. Workers have reported that the addition of nicotinamide adenine dinucleotide (NAD^+) or glutathione prevented such loss of activity. Others claim that storage for a few days at 4 °C is perfectly adequate and this seems to be one of those situations where laboratories must decide on their own policy as a result of information which is available from the scientific literature and personal investigations.

In some instances loss of activity may be reversed. For example, creatine kinase readily loses its activity as a result of sulphydryl group oxidation at the active site of the enzyme. Activity may be at least partially restored by the addition of sulphydryl compounds such as N-acetylcysteine or thioglycerol.

The physical characteristics of plasma and serum specimens can be affected by storage. The analytical problems caused by the formation of small fibrin clots in plasma samples has been described (3.1.2). After standing in the refrigerator overnight chylomicrons tend to separate out giving two distinct phases within the serum specimen. When serum that has been stored in the deep-freeze is thawed a watery layer can be clearly seen on the surface. These samples must be mixed well before the analyses are performed.

In view of the problems of the variable lability of constituents in blood, plasma or serum it is a sound policy to perform analyses on specimens with the very minimum of delay following their collection. Microbial degradation is not normally a problem. It can be a source of error in field surveys when specimens must be transported over long distances and may arrive at the laboratory several days after they were collected. In such cases antimicrobial compounds such as sodium ethylmercurithiosalicylate at 0.2–0.5 mg cm^{-3} have been added to specimens. It would seem prudent to investigate thoroughly the effects of such substances on the intended analyses before using them routinely.

3.5. HAZARDS ASSOCIATED WITH BLOOD SPECIMENS

Blood is not corrosive or dangerous to work with in any physical sense. However all specimens should be regarded as being potentially infective and the consequences of the more serious infections which can be transmitted by blood and blood products must be well understood by laboratory workers. Staff should be aware of the main routes of transmission of infection and appreciate the need for working practices which are designed to minimise the hazard. According to the Advisory Committee on Dangerous Pathogens (United Kingdom) there are three relevant routes of transmission of infection via blood specimens. These are:

- percutaneous transmission by needle or other sharp objects;

- percutaneous transmission by contamination of cuts, scratches, abrasions, burns etc;

- contamination of mucosal surfaces of the mouth, nose or eyes. Direct transmission is possible by mouth pipetting or from splashes. Direct contact with a contaminated surface and then from hand-to-hand, or hand-to-eye is less likely but possible nonetheless.

The production of true aerosols (particles less than 10 μm in diameter) from blood or serum is not likely, but splashing and the transient presence of large droplets from specimens can occur and may be inhaled by workers nearby.

The two diseases which are of particular concern in laboratories which process large numbers of blood samples are hepatitis and AIDS. Statistics on hepatitis B infections in laboratory workers may not yet be complete and there is little information on the incidence of the infection in other occupational groups with which to draw comparisons. AIDS is of much more recent concern, but again the true risk cannot be evaluated due to lack of meaningful data.

It is not appropriate in this text to discuss the identification and testing of patients in 'high risk categories' for either hepatitis or AIDS or to describe procedures required for the safe handling of blood, serum or other clinical samples. These topics are under continual review. Persons who are responsible for safety policy in laboratories must design working practices based on information from current publications of the national and international advisory bodies.

Summary

Venous blood is the most commonly collected blood specimen followed by capillary blood with arterial blood a very poor third. Proper collection technique is essential if specimens are to reflect accurately the composition of circulating blood.

The choice of whole blood, plasma or serum for biochemical assays depends upon a number of factors such as the distribution of constituents between cells and plasma, the effects of intracellular substances or anticoagulants on reactions used for the assays and any changes which can alter the composition of samples after they have been collected.

All blood and blood products must be regarded as potentially infective and be handled in a way which minimises this hazard.

Objectives

You should now be able to:

- discuss the advantages and disadvantages of using venous, capillary and arterial blood for biochemical analyses;

- explain the necessity for proper blood collecting techniques;

- discuss the factors which influence the choice of whole blood, plasma or serum as analytical samples;

- describe the action and applications of the more commonly used anticoagulants;

- identify the necessity for accurate patient and specimen identification;

- describe the changes which may occur in blood specimens after collection;

- describe the hazards of infection from blood specimens.

4. General Points on Biopsy

Overview

Some common types of biopsy specimens used in clinical investigations are considered, together with the methods employed for their removal and the problems involved in their analysis.

Blood, as the most important type of specimen, and fetal materials, because of their specialised nature, are dealt with elsewhere.

The term biopsy is one of those ill-defined words with a number of shades of meaning. It really refers to the removal of a sample of human tissue for the purpose of clinical investigation, and as such should strictly include the taking of liquid samples including blood and urine. However such sampling is so easy in contrast to tissue and organ sampling that the term is usually restricted to these latter types. Tissue biopsy specimens from living patients are extensively used in histopathological work including the investigation of cancers, and are much less frequently used in purely chemical studies although on occasion they are of great importance.

Perhaps the most common tissue biopsies are those of the liver, gut, lung, bone marrow, skin, various fetal tissues, and pleural and pericardial membranes. Apart from blood, urine and faeces, the more commonly requested fluids are cerebrospinal fluid, lymph, amniotic fluid from around the fetus, aspirates of the gut contents, pleural fluid from the lungs, and synovial fluid from the joints.

The most significant general problems with obtaining the biopsy type of sample are those associated with patient acceptability and preparation, since anaesthesia is usually required and some discomfort or pain experienced.

SAQ 4.0a

What do you think are the more significant scientific problems with obtaining biopsy specimens?

4.1. ORGAN SAMPLING

It is theoretically possible to take samples of virtually any organ of the body but in reality the practical problems of obtaining them, and the consequences to the patient have meant that many body organs are infrequently or indeed never sampled.

4.1.1. Liver Samples

Among those that are sampled, skin and liver are perhaps most commonly requested, but only the latter is used to any extent for chemical investigations. Liver biopsy was first performed by Paul Ehrlich in 1883 as part of a study on liver glycogen and the technique has grown so that the Royal Free Hospital, London carried out more than 6300 such biopsies between 1975 and 1985. Most commonly the sample is requested for investigations of diffuse parenchymal disorders, eg chronic hepatitis, cirrhosis or alcoholic liver disease and while histological investigations are perhaps most significant, many chemical measurements are undertaken in these studies. In addition a range of chemical investigations are made into inherited diseases using these specimens. Studies of the eight or so glycogenoses require the microanalysis of the enzymes involved in glycogen metabolism, and of the detailed branching structure of glycogen itself. Confirmation of Wilson's disease can involve a measurement of copper levels using neutron activation analysis of the liver sample.

In principle sample taking is relatively straightforward. The patient should be mentally relaxed, locally anaesthetised and then a syringe loaded with about 3 cm^3 of sterile saline is inserted through the skin (usually between the ribs) in the direction of the liver. About 2 cm^3 of solution are injected to clear the needle of skin and tissue fragments before the liver is penetrated. A core of 1–4 cm length (10–50 mg) can be taken by aspiration (sucking), usually without the need to rotate the needle which could possibly damage the cells. Among the needles most commonly used is the Menghini type (Fig. 4.1a) in which sampling is aided by a depth gauge, and a small rod contained within the needle dampens the pressure changes and decreases fragmentation of the sample. Failures occur most commonly when the liver is tough, eg in cirrhosis, or is small or displaced, eg in emphysema, in which case it can be physically missed during the sampling. In common with the biopsy sampling of most organs, haemorrhage is the most common (although infrequent) complication, and consequently patients undertake a closely observed rest period after the operation. Biopsies are in fact not usually performed at all if there is a history or evidence of a bleeding disorder in the patient.

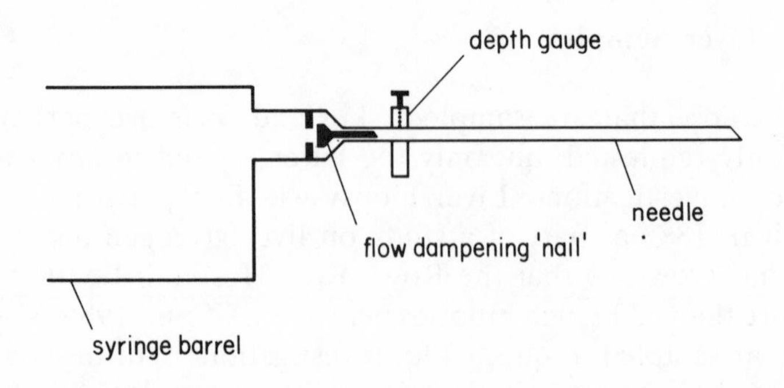

Fig. 4.1a. *Menghini-type sampling needle*

SAQ 4.1a

There are three major types of jaundice:

(*i*) pre-hepatic, in which excessive red blood cell breakdown occurs and sufficient haemoglobin is produced to overcome the ability of the liver to remove it for breakdown and excretion in the bile;

(*ii*) hepatic, in which the liver cells are defective and cannot convert the haemoglobin properly or at an adequate rate into the derivatives excreted in the bile;

(*iii*) post-hepatic, in which the bile is unable to escape from the liver usually due to a blocked bile duct.

When other investigations indicate that the patient has pre-hepatic jaundice it is not really worthwhile carrying out a liver biopsy. Why do you think this is?

SAQ 4.1b

> Bile is a relatively toxic material and hence it is sometimes useful to investigate a liver biopsy when studying a serious or prolonged case of post-hepatic jaundice. What is the justification for this?

4.1.2. Lung Samples

Lung biopsies are frequently requested in infections and are taken either by an aspiration or cutting needle. Apart from cytological and micro-biological investigations, lactate dehydrogenase assays are sometimes performed to monitor exudation processes, and amylase activities measured if a ruptured oesophagus is suspected.

4.1.3. Muscle Samples

Biopsies from the major skeletal muscles are easily obtained and quite frequently requested. Using a local anaesthetic three cylinders of 1 mm × 2 cm are taken and sent for structural study by both light and electron microscopy, and for histochemical investigations of enzyme distribution.

4.1.4. Gut Samples

Gut samples, especially of jejunum and ileum are used for the investigation of gut function, eg digestive enzyme deficiency in certain inherited diseases. Among the latter is lactose intolerance, commonly found in negro populations, which can be identified by demonstrating the absence of the enzyme lactase in jejunum biopsies. Failure to take up radioactive amino acids into gut wall cells, or to esterify fatty acids into triacyglycerols, are other possible investigations carried out in diagnosing digestive problems.

4.2. LIQUID SAMPLES

4.2.1. Saliva

Saliva is of course a well-known oral fluid with lubricant, weakly biostatic and digestive roles. It has a flow rate varying from virtually zero to 3 to 4 cm^3 min^{-1}, and a definite circadian rhythm in composition as well as flow rate. Being the most readily accessible of body fluids has led to some interest in its use in diagnosis but in practice relatively few such uses have been identified. Among those that have, are studies of electrolyte changes, eg in aldosteronism, or interestingly, in the fertile pre-ovulatory period of the menstrual cycle which can act as a useful signal to couples wishing to conceive. Measurements of the amylase and bicarbonate content are said by some workers to be a better indication of pancreatic function than the secretin/pancreozymin hormone stimulation test. Blood groups particularly of the H and Lewis types can be identified by agglutination reactions, which is of considerable value in forensic work, and lastly saliva accumulates many compounds, eg drugs such as phenytoin, in a non-protein bound form. This not only makes assays rather easier but also makes them more meaningful since it is the unbound form which is physiologically active. Some argument exists however over the correlation of the salivary and serum values since the pH difference will effect the bound/unbound equilibrium.

Collection of saliva does not involve merely spitting into a bowl since such a sample would be contaminated with mouth, oesophageal and bronchial lining material and fluids. In practice specially shaped cups are fitted over the different salivary glands and saliva collected by gravity or gentle suction. The flow of saliva is usually stimulated by the chewing of some plastic or rubber device although this is said to cause the loss of some lipophilic material by absorption. The use of chewing gum in some American institutions has been largely abandoned due to the physiological effects of the sugar and flavouring content.

4.2.2. Gastric Contents

While satisfactory patient co-operation is important it is relatively easy to insert a tube via the nose into the gut, although accurate positioning of the collecting end of the tube along the length of the gut can be troublesome. Originally the tube used was marked with appropriate distances, eg 45 cm which is an average value for the distance from the teeth to the stomach opening and 75 cm which would reach to the pancreas. The advent of fibre-optic devices has greatly simplified this problem and fluoroscope control of location is now routine. It is usual to collect and generally discard the existing (residual) fluid following the insertion of the tube and then collect the fasting (or resting) material for the next hour or so.

For some investigations stimulants are used, not just to increase the volume collected, but as part of an assessment of the gut physiology of the patient. Thus pentagastrin, which is a peptide related to the hormonally active part of the molecule of gastrin but which does not induce immunological and other adverse reactions when given experimentally, is used in an assessment of the ability of the stomach to secrete the digestive fluids following hormonal stimulation. Studies of the function, or infection or bleeding into any part of the upper gut are the main justifications for this type of investigation, and if necessary samples can be taken from as far down the gut as the duodenum in order to obtain pancreatic secretions or bile.

4.2.3. Faeces

A typical adult produces between 50 and 200 g of faeces per day although in certain alimentary conditions resulting in poor absorption this can rise to over a kilogramme. Investigations of clinical conditions by faecal examination are not popular for obvious reasons, and it is perhaps fortunate that comparatively little useful information can be obtained from them compared with that from blood specimens.

The unpleasant nature of faecal specimens (particularly when the alimentary system is defective or infected) means that collection and examination techniques need to be designed to minimise exposure of staff to them. Accordingly a straightforward collection will involve the rapid transfer of the total faecal deposit to a sealable bucket, or the removal of a portion to a waxed, water-proof, pre-weighed carton using a spatula similar to those provided with ice-cream tubs some years ago. Transport to and examination within the laboratory can then be done more pleasantly although every laboratory has its horror stories of accidents, usually to the more noxious specimens!

Patient preparation normally presents few problems. Standardised diets are used on occasions if comparative results between individuals, or for a given individual over a period of time are of interest. Examination for the so-called occult (or hidden) blood, arising due to blood leakage into the gut, requires a high fibre diet, a reduction in the consumption of red meat and some particular vegetables (eg turnips and horse-radish). Mild gut irritants such as vitamin C and aspirin need also to be excluded.

There is no question that more problems exist over the collection and study of faecal specimens than most other common biological samples. The simple fact that it cannot usually be delivered on demand immediately implies that significant patient/staff collaboration, time, effort, and hence in the end, money, will be involved in its collection.

SAQ 4.2a

Single faecal samples have a number of disadvantages from an analytical point of view. Consider what they might be.

As a consequence of these problems it is common to pool together deposits obtained over a longer period, say 1 to 3 or 4 days, or to take several consecutive specimens for analysis.

SAQ 4.2b

When collecting these longer term specimens a problem exists in correlating faecal deliveries with the chosen time period for analysis. Orally taken dyes are frequently used to identify the beginning and end of the chosen collection period. How can this be done?

While the use of different dyes at the two ends of the chosen time period is quite common a rather better approach is to dose with some non-absorbed, measurable material (X) at intervals during it. All the faeces are collected, homogenised, and the total volume and the concentration of X then determined, along with the parameters of clinical interest. Arithmetical calculations will provide the number of days or fractions of days of faecal output represented by the aliquot of homogenate being studied. Chromium sesquioxide (Cr_2O_3) is often given in 3 × 500 mg capsules over a 24 h period. The concentration of Cr can be determined by atomic absorption spectroscopy and the volume of faecal homogenate corresponding to 1.5 g of Cr_2O_3 therefore constitutes 1 day's output of faeces by the patient.

SAQ 4.2c

A faecal deposit of 200 g is homogenised in 1 dm^3 of water and produces 1250 cm^3 of homogenate. A 750 cm^3 volume of homogenate is investigated and has a Cr_2O_3 concentration of 1 mg cm^{-3}. Assuming the patient was supplied with 1.5 g of Cr_2O_3 as described above, what proportion of a days output of faeces does the 750 cm^3 represent?

Contamination and intrinsic variability are more significant with faecal samples than with most others and are perhaps the main reasons for the assays of faeces being restricted to specialised tests only.

Contamination with urine, blood and toilet preparations (which often contain oxidizing agents) is particularly easy for females, and with enema materials in both sexes. Barium salts in the latter produce significant interference with faecal fat estimations, which are the most commonly requested analysis of faeces.

The action of bacteria on faecal specimens both during the later stages of their transit through the gut and after collection produces significant, but more importantly, variable changes in the composition of the specimen. Preservation is attempted by either refrigeration, drying and/or formalin treatment.

As a consequence of these changes it is important to note the colour and appearance of the faeces upon collection since this can give important clues as to illness but which can disappear or change nature on standing. Thus bilirubin and biliverdin produce yellow/green colours, whereas significant haemolysis yields increased amounts of urobilin and very dark faeces. Urobilin production continues on standing so that the surface of the faeces darkens on exposure to light and air. Fresh blood due to lesions in the rectum or anus leaves visible streaks on the surface of the deposit. Faecal colour can also be affected by diet and malabsorption syndromes.

4.2.4. Urine

Urine Production. The body is a complex chemical machine that has to undertake a balancing act involving the taking in of food, the removal from it of required materials and then the carrying out of its life's activities (movement, temperature maintenance, repair, growth etc). As a consequence of these activities wastes must be eliminated, including materials in the diet which are not required by the body, wastes generated by the body, and to some extent important materials lost during normal metabolic 'turnover'. The lungs, liver, lower gut and especially the kidneys have major roles in this excretion. The kidneys constantly remove metabolic wastes, drugs and other foreign compounds (collectively called xenobiotics), salts, acidic and basic substances, and excess water, for excretion in the urine. Each kidney has about a million tubular systems (nephrons) with the job of filtering blood, adjusting the concentrations of the filtrate components by selective excretion and re-absorption in line with the body's needs at any given time, and then excreting the remainder to the exterior as urine.

An important point is that very careful control of secretion and re-absorption is necessary, to maintain the composition of most components of the body within relatively narrow limits. Since it is acting as a 'balancing' device in this way, the composition of urine can vary quite widely even in health. This variation occurs between individuals of course, but also for a given individual from time to time. As an example of this the urine volume varies from 800–2000 cm^3 day^{-1}, with an average of 1200 to 1500 cm^3. However a complexity is that while the volume in children is similar, ranging from 300–1500 cm^3, their output per kilogram of bodyweight is 3–4 times that of an adult. The interpretation of data which show a relationship to body weight thus becomes more difficult, especially if the patient is adolescent and moving between the two extremes.

The factors that affect urine volume and composition can usefully be divided into:

- dietary effects, eg an excessive or insufficient intake of water, salts etc and

- the effects of physiological processes, eg physical activity, extensive body repair etc.

In disease the urine can vary due to:

- primary effects in the kidney, eg the impairment of re-absorption of amino acids by heavy metal poisoning of the transport systems, or a failure to re-absorb water due to ineffective vasopressin action,

- secondary effects due to a condition originating elsewhere, eg the action of bacterial toxins on various aspects of nephron function.

Urine Sample Collection. When single collections are made the first voiding in the morning is preferred since it is more concentrated, less variable and its lower pH helps to preserve a number of analytes. The fact that it will have been produced after overnight fasting makes the analyses more meaningful for some investigations since the results will be less affected by recent meals. Additionally its longer incubation time in the body will make any bacterial contamination, due to patient infection, more obvious and allow the specimen to be discarded.

However a significant problem with urine production is that a number of materials are secreted or excreted unevenly through the day (ie they have a circadian rhythm); thus the catecholamines and bilirubin excreted in the urine have their highest concentrations in the late afternoon and if a single collection is envisaged, then 4 pm is perhaps a better time than 8 am. Another source of difficulty is that a greater volume of urine is produced during the waking compared with the sleeping periods; the ratio commonly being 2 : 1 but may be as high as 3 : 1.

SAQ 4.2d

Even if a material is excreted in urine evenly throughout the day there will be a variation in concentration due to differences in volume produced between day and night. Would an early morning or early evening specimen be the more dilute?

An apparently straightforward way of overcoming this problem is to collect complete 24 h voidings of urine as pooled specimens. The standard technique for doing this is to discard the 0800 h specimen on a given day and collect all the specimens for the next 24 h period including the 0800 h specimen on the following day.

A number of problems and errors arise with such collections and in advance of the following text you might like to consider what some of these are.

The following are perhaps the more important of these problems and they arise out of the facts that urine is an excellent growth medium for micro-organisms, and that logistical difficulties exist over the actual collection at home and even in a hospital ward:

- contaminated containers might be used;
- specimens might be biologically contaminated;
- inappropriate or low concentration preservatives might be used;
- poor storage conditions might exist;
- specimen might be lost;
- the time period for the collection might be incorrect.

Contamination with other materials is relatively easy for urine; toilet preparations and blood being obvious examples. Particularly in a home environment the initial collection in any convenient container before decanting into the one supplied by the hospital can lead to contamination with almost anything and most laboratories can supply tales of urines containing items from fish-bones to hair-grips!

Urine changes significantly on keeping, primarily due to the action of bacteria. If voided into clean containers and kept cool, these changes are not very significant over a 1–2 day period, but we have all met the strong ammoniacal smell arising from badly cleaned toilets which is good witness to the action of bacteria on the nitrogenous compounds (principally urea) in urine. Even with strict cleanliness 24 h collections from patients who are catheterised can present problems of this type if the urine is kept standing in the drainage bag for any significant length of time.

Other changes which occur in urine are the precipitation of uric acid and urate salts as the urine cools, and the oxidation of some materials, especially bilinogens and bilirubin when urine is exposed to light and air. Thirty minutes exposure to room temperature and light will cause the virtual disappearance of bilirubin in a normal urine. A number of compounds form melanin type pigments after standing for 12–24 h, giving brown or black urines. Melanogens from malignant melanomas, and homogentisic acid produced in the inherited disease of alkaptonuria are examples of this process.

The object of preservation is therefore to decrease microbial action, and minimise the spontaneous or microbial degradation of solutes, large molecules and particles, eg haemoglobin and voided cells.

Among the common preservatives are strong acids eg 10 cm^3 of concentrated HCl which is effective, if a little dangerous. It is also unsuitable for some assays, eg of steroids and bilirubin. Thymol and to a lesser extent, formalin are also used but the latter interferes with copper reduction tests for urinary glucose. In any case many organic compounds have significant problems of miscibility with the sample or of interference with analytical reactions.

Obtaining a proper 24 h voiding of urine is in principle easy but in practice more difficult due largely to the domestic problems of remembering to collect all the urine and particularly if the patient is not hospitalised, carrying a sufficiently large container with them throughout the working day. Some individuals find it difficult, or are unaware of the need to empty their bladder fully and loss of specimen while defaecating is particularly common. The situation is undoubtedly more difficult for females than males, due to the greater difficulty of collection and greater likelihood of embarrassment. Excess specimen can be produced if both early morning specimens are included. Patients and hospital staff need to be thoroughly advised about these problems.

For many years it has been said that the measurement of creatinine concentration in the urine gives a reasonably good indication of the extent of a 24 h voiding since most creatinine is of endogenous origin, and hence, except on very high meat diets, the output is directly proportional to the muscle mass and is therefore constant for a given individual. Also, for individuals of average build, the variation within the population for a 24 h output is relatively small. Creatinine output should therefore be useful as a check on the completeness of a 24 h voiding. As with most other scientific notions, detailed studies have cast some doubt on the general validity of this assumption but it is still commonly employed.

When actual studies on kidney function are involved, a frequent approach is to use 'challenge' tests where the body is dosed in some way and its response elicited, usually by parallel measurements of urine and blood concentrations. Patient preparation for these tests can be involved as the following protocol for the so-called inulin clearance test will show. In this case the patient is to:

- have no exercise on the day in question;
- fast for at least 4 hours;
- drink about a litre of water an hour before the test begins;
- be intravenously infused with a dose of 500 cm^3 of 1.5% inulin solution;
- supply a urine specimen 30 minutes later, and three more at 20 minute intervals;
- supply 2 blood specimens at mid-points between the urine specimens.

It is perhaps with the hormone assays in urine that patient pretreatment is most significant with a wide range of factors such as diet, physical exercise, and emotional changes affecting urinary hormone levels. Aldosterone levels are affected by sodium-rich foods especially processed meat products (bacon, stock cubes, sauces etc), and anti-hypertensive drugs which increase sodium retention. Both cause a decrease in aldosterone level. Diuretic drugs on the other hand increase aldosterone output as the body strives to regulate its water and sodium levels in the face of the increased urinary water loss due to the drug.

Samples collected for cytological investigation, eg in the study of cancer and general inflammation of the urinogenital system present their own problems, primarily as a result of the lability of the voided cells and the difficulty of finding preservatives that will not adversely affect them. A typical procedure for the collection for such investigations would be to clean thoroughly the external surfaces of the urinogenital apparatus and collect 100–300 cm^3 of urine from the mid-stream of voiding. While the overnight urine is not normally used it is considered useful for cytological work since the significant time lag from the previous voiding will have allowed the build-up of cells in the urine. Catheterisation reduces the risk of contamination by micro-organisms or surface cells but currently carries a risk of urinogenital infection in 1–2% of cases. Samples must obviously be transported to the laboratory as soon as possible and examined less than 1 hour after collection or less than 18 h if refrigerated.

Sample collection from children and especially babies is fraught with difficulties and while special systems for taping collecting devices to the urinary outlets have been marketed, in many cases it is better to aspirate the bladder directly by a syringe inserted through the groin wall, although this is a procedure requiring significant surgical skill.

Calculi and other Solid Matter. The urine can contain a variety of solid materials of considerable diagnostic significance. Among these are cells shed from different parts of the urinogenital system, casts which are relatively small chemical aggregates and calculi (stones) which are inorganic or organic crystalline deposits varying in size from pin-head to hen's egg. Casts and calculi tend to accumulate due to changes in body metabolism leading to excess excretion of various organic and inorganic materials and/or changes in pH or flow rate which increase the ease of precipitation or aggregation of the chemicals.

In general a single mid-morning (but not first voiding) specimen collected from the mid-stream of the delivery is adequate for these investigations. Owing to their nature it is essential that a clean container is used for the collection and, bearing in mind the instability of the cells and casts, that a rapid examination is undertaken. In some hospitals nursing staff and patients are asked to strain the urine through muslin or a sieve but in the majority of cases the urine is sent directly to the laboratory for analysis. In some situations specimen collection can be more troublesome, large calculi developing in the bladder usually require surgical removal and identification of the site within the urinogenital system where cells are being shed may require the insertion of cannulae to specific locations.

It is well to note that most, if not all, tubular 'flow-through' systems within the body can develop these types of solid obstruction with the urinogenital system, gall-bladder, pancreas and salivary glands being the most well known.

4.2.5. Cerebrospinal Fluid

The central nervous system of the body (brain and spinal cord) is encased by three complex membranous structures called meninges, and a special fluid, the cerebrospinal fluid (CSF) accumulates between the inner and middle of these in the so-called sub-arachnoid space. It is generally considered to have the roles of cushioning the central nervous system (CNS) against physical shock and to some extent supplying nutrients and removing waste products.

The CSF is produced by active secretion from the central nervous system vasculature at a rate of 500 cm^3 day^{-1}, flows over the CNS and is ultimately reabsorbed. With a total volume of only 125 cm^3, the excess production gives rise to a hydrostatic pressure equivalent to 13 cm water (12.75 N m^{-2}) which generates an unusual sampling problem. If too large a sample volume is removed the pressure drop allows blood vessels to expand which can produce severe headaches in the patient.

Major changes in CSF composition occur as a result of infection, inflammation, tumours, tissue degeneration, vascular changes and trauma (eg accident and physical damage). Hence the analysis of CSF is often of some interest.

Samples can be taken from the large cavities of the brain, but the potential hazards involved mean that, in general, the lower spine is preferred since a readily accessible area exists in which the CSF cavity is present below the point where the spinal nerve cord itself has ended. The decrease in hazard is sufficient to outweigh the fact that the composition of the CSF is slightly different from that in the area of the brain, protein concentrations for example are only 0.1–0.25 g dm^{-3} for brain CSF, but 0.15–0.45 g dm^{-3} for the lumbar CSF.

The sample is collected by a needle inserted between the lumbar vertebrae, hence the popular descriptive term 'lumbar puncture'. At the time of collection it is usual to measure the CSF pressure since physical obstruction, eg by tumours, will decrease this, whereas meningitis and other infections will increase it. Samples are used for both microbiological and chemical investigations which unfortunately have different preservation, transport and storage requirements. Since in general the former are more important, samples are taken with these in mind and sent first to the microbiology laboratory, or subdivided and sent separately to the two laboratories.

SAQ 4.2e

Bearing in mind the procedure involved in collecting CSF, with what might the collected fluid be contaminated, and what effects might these contaminants have on the clinical value and reliability of the investigation?

4.2.6. Synovial Fluid

The synovial fluid is a specialised liquid mixture contained within membranous sacs situated between the bones at the major joints of the skeleton, eg the knees and elbows. Its primary role is to allow the joint to move freely by acting as a lubricant, although in fact it has quite complex properties due to the different needs of the joints when placed under different intensities and angles of force. A range of quite common clinical conditions exist in which the fluid is reduced in quantity, altered in composition or is the site of inflammation or infection.

While sterile techniques and anticoagulants are required, sampling of synovial fluid from the large joints is relatively easy and therefore quite common. In general about 10 cm^3 of sample are removed and dispensed into three separate containers for biochemical, cytological and microbiological investigations. Samples for the former usually omit anticoagulants.

In addition to studies of cell number and type and the examination for clots produced by certain types of inflammation, a range of chemical investigations are frequently undertaken particularly in the investigation of rheumatoid arthritis. Here inflammation can result in an increase in the difference in glucose concentration between fluid and blood and the actual value may fall to half normal. Complement proteins involved in the immunological processes decrease, possibly due to their consumption in the antigen/antibody reactions of arthritis, but about 80% of individuals with rheumatoid arthritis develop a measurable anti-gamma globulin type M antibody (the 'rheumatoid factor'). As another example of the use of this fluid the chemical or visual demonstration of significant uric acid accumulation is a useful confirmation of gout.

4.2.7. Sweat

Sweat is a body fluid that is more easily accessible than most others but unfortunately is only diagnostically useful in the investigation of a relatively small number of conditions. It is not necessary to resort to vigorous physical activity in order to collect it, since electrical stimulation of the skin will induce the penetration of sweat-stimulating drugs such pilocarpine. As a consequence quite reasonable amounts of sweat are released and can be collected. The sweat is usually collected in an absorbant pad from which it can later be washed; precautions to reduce bacterial metabolism of its contents are not usually undertaken. The main clinical area in which sweat analyses has found some application is in the investigation of inherited diseases, the most noteworthy being the identification of raised sodium chloride levels in cystic fibrosis in children. While the interpretation of the results can be troublesome, the measurement is important since the disease is ultimately a fatal one and unfortunately is among the most common of the inherited diseases.

SAQ 4.2f

Use your knowledge of general chemical analysis to suggest some of the different ways in which the increased level of sodium chloride in the sweat of individuals with cystic fibrosis might be shown, at least in a semi-quantitative way.

4.2.8. Pleural Fluid

The so-called 'pleural cavity' surrounding the lungs is in fact really a potential cavity since it usually contains only sufficient fluid (say 10 cm^3) to act as a lubricating surface during lung expansion. However under some circumstances this can rise to over 1 dm^3, when needle aspiration is required for patient relief as well as for analysis. Apart from investigations for infection and malignancy, measurements of pH and gas content are common. A ruptured oesophagus produces pH values <6, infection values of 6–7, and malignant effusions frequently have values of >7.4.

Summary

Some common types of tissue and liquid biopsy specimens and the methods used for their removal are described. Some of the problems in obtaining a representative sample and maintaining its integrity prior to and during the investigations are discussed together with a number of other problems arising from the particular nature of individual specimens.

Objectives

You should now be able to:

- list the major types of human biopsy specimen used in diagnosis;
- describe the processes involved in obtaining a representative sample of these materials;
- discuss in some detail the various problems involved in sampling, and in maintaining the biological integrity of the material prior to and during the investigations.

5. Fetal Samples

Overview

The investigation of a fetus to evaluate its age, determine its sex, or identify a wide range of clinical conditions (many of which are hereditary) is a very important and rapidly developing area of diagnosis. This part is primarily concerned with the techniques for obtaining samples for these investigations but some indication of the analytical procedures involved is also included.

5.1. AMNIOCENTESIS

5.1.1. The Basic Technique

The principle of this sampling technique is to penetrate the amniotic cavity in order to obtain samples of fetal material, which can then be used to study the state of the fetus, or occasionally that of the mother. Investigations usually involve chemical studies on the amniotic fluid itself, cytological or chemical studies on cells released from the fetus and suspended in the amniotic fluid, or similar studies on blood or tissue samples taken from the fetus or placenta directly. A range of variations on the technique are available but in general they involve the insertion of a hollow needle (Fig. 4.1a) to allow the withdrawal of fluid and/or tissue samples. Of the two most obvious routes of entry the transabdominal approach is much more common since a vaginal route of entry has been shown to increase the risk of fetal abortion quite significantly. Originally the technique was performed under a general anaesthetic but is is now more common practice to use a local anaesthetic. Samples of about 20 cm^3 volume are usually taken to provide enough material for a duplication of cell cultures and a range of biochemical investigations. With care cells present in the withdrawn fluid will survive at least 2–3 days which allows samples to be dispatched to other laboratories for specialist investigations or merely to duplicate the investigations as a safeguard. The majority of hospitals carrying out this technique begin with an ultrasound scan of the mother's abdomen since this produces a visual image of the fetus on a TV monitor.

SAQ 5.1a

An ultrasound scan of the mother's abdomen is extremely useful to the surgical team for a number of reasons. Bearing in mind what is actually being done to the mother, what do you think are the more straightforward of these benefits?

It is necessary to wait until at least 14–16 weeks into the pregnancy before undertaking amniocentesis so that there is sufficient fluid to allow the removal of a sample without serious effects on the fetus. In general the technique is carried out in the second trimester (ie one-third period) of pregnancy, generally between 16 and 18 weeks.

Considerable concern has been expressed over the risks of the amniocentesis techniques and it is unfortunate that precise data are difficult to collect. It is generally accepted that the risk to the mother is negligible; at an international conference in 1979 only one maternal death was identified out of a total of 20 000 cases known, and the uterine infection rate was considered to be only about 0.1%. More insidiously however it, was confirmed that a significant number of mothers suffered mental distress, probably because the primary reason for the investigation was to study abnormalities and to allow a consideration of therapeutic abortion.

There is a somewhat greater risk to the fetus. The normal level of spontaneous abortion is probably about 1–2% and the technique increases that by about another 1%. Unfortunately many reports indicate that there are significant non-fatal changes in the fetus. An increase in respiratory and musculo-skeletal disorders seen at birth has been suggested, and many babies seem to be of low birth weight, possibly because of a 'starvation' effect due to the removal of the fluid sample.

5.1.2. Principal Types of Investigation Using Amniotic Fluid

(*a*) Chemical Assays. Chemicals of a wide range of types from simple organics, eg creatinine, non-enzyme proteins such as α-fetoprotein (AFP), enzymes, and hormones, can be measured in the fluid obtained by this technique.

(*b*) Cell Investigations. Cells can be studied in the original amniotic fluid (ie without culture), or after short term or long term culture.

5.1.3. Technical and Sampling Problems

A major problem can be the contamination of the amniotic fluid sample with fetal and maternal blood. For investigations of α-fetoprotein (AFP) level (see later) this is not of great significance since it is present at only very low levels in maternal blood, but for investigations of the lipid composition of fetal lung surfactant it is of great importance that the sample is not contaminated with blood, meconium or vaginal mucus.

SAQ 5.1b

> Why do you think this point concerning lung surfactant investigations is important, and can you imagine (in very simple terms) how the procedure might be modified to minimise it?

The presence of more than one fetus is another potential problem since this is likely to alter the concentration of materials in the fluid.

Perhaps the largest number, and most significant problems, are encountered with those investigations involving cell culture. Actual culture failure rate is quite low with modern techniques and experience, (2–5% for most centres), and the practice of duplicating cultures in different centres makes this a relatively insignificant problem.

Amniotic fluid contains a wide range of cell types many of which are of potential value because of their specialised nature. Unfortunately many of them frequently fail to grow due to a failure to attach to the culture vessel surface or for other reasons, or may be overgrown and crowded out by more active cells in the mixed population. It is noticeable that early in the growth of the cultures the cell population is a mixture of various epithelial and fibroblast-like types, whereas later on in the growth the fibroblasts begin to predominate. Much potentially useful information is lost because of these effects.

A significant number of cells in the fluid can be expected to originate from the mother and indeed both fluid and cells are present in empty amniotic sacs and in those with ova which fail or cease to develop. In a normal pregnancy large numbers of cells appear at about 14 weeks and while the number increases steadily the percentage that are viable actually falls. By about 18 weeks the fluid will generally contain about 5×10^4 cells cm^{-3} and following this time the number in female fetuses rises substantially above that for males due to the shedding of large numbers of cells from the fetal vagina. Cell number is too variable a parameter to be of any diagnostic significance.

SAQ 5.1c

Cellular enzyme activities vary with a number of factors and you could consider what you think some of the factors contributing to this variation in cellular enzyme level might be. While you may be unfamiliar with the biological processes employed, just bear in mind that basically they involve placing the cells in a suitable growth medium, allowing them to grow, divide and multiply with periodic replenishment of the medium, until sufficient cells can be harvested for analysis.

SAQ 5.1c

Cells present in the original amniotic fluid and in the primary cultures are much more variable in their enzymic activities than those in subsequent cultures. One interesting report discussed two separate amniocentesis samples from the same fetus, with the cells being cultured through the same number of passages and yielding virtually a 100% difference in acid hydrolase enzyme activity. These problems can be somewhat reduced by a study of enzyme ratios in the cell extracts rather than the individual enzymes in isolation. Nonetheless a major problem lies in the determination of reference ranges for particular chemicals and enzymes and it is essential that each laboratory determines these for its own procedures.

Attempts to speed up the growth rate of cells to allow earlier diagnosis have largely been unsuccessful, and instead substantial effort has been directed towards improving the sensitivity of the assay techniques by using radioactive or fluorescent substrates, labelled antibodies etc. These 'micro-analytical' techniques have reduced the cell requirements from several million cells to several thousands, and by allowing assays to be performed on sample volumes as low as 5–10 μl provide much earlier diagnosis of many conditions. Significant cost, humanitarian and even legal advantages accrue from this.

5.1.4. Fetoscopy and Fetal Tissue Sampling

This technique is a modification of amniocentesis in which a fibre-optic device (a fetoscope) is inserted down the tube used to penetrate the amniotic cavity thus giving direct visualisation of the fetus. It has enabled surgeons to take samples from the fetus itself without very significant hazard to fetus or mother.

SAQ 5.1d There are perhaps two main applications of the fetoscopy technique; in very simple terms what do you think they are?

Principal Types of Fetal Sample

(*a*) Blood Samples. Blood samples can be obtained from the chorionic plate or a vessel in the umbilical cord. The latter is preferred since it gives a purer fetal blood specimen and for anatomical reasons the puncture re-seals more rapidly after the withdrawal of the needle. The presence of fetal red cells can be confirmed by Coulter cell size analysis since the fetal cells have a volume of 135 fl (1.35×10^{-16} m^3) whereas the maternal ones are only 90 fl (9.0×10^{-15} m^3). Alternatively the Kleihauer assay will confirm the presence of the fetal haemoglobin (hbF). Fetal red cells can be isolated by agglutination or lysis of maternal cells, or by density gradient centrifugation, thus allowing a study of their characteristics to be undertaken.

(*b*) Skin Biopsy. Samples of a few square mm can be taken from the back, thighs or scalp of the fetus by special forceps without any apparent ill effects. These allow biochemical investigations, and light and electron microscopic studies of chromosome and whole cell structure to be undertaken.

(*c*) Liver Biopsy. A number of important inherited diseases can only be identified from changes occurring in specific cells as they are not reflected in chemical changes within the amniotic fluid, nor are those cell types released into the fluid. Ornithinecarbamoyl transferase deficiency which results in a failure to convert ammonium ions to urea in the liver is one of these. Liver tissue specimens can be obtained by direct puncture with a wide bored needle through the fetal abdomen.

5.1.5. Chorionic Plate Biopsy

For reasons discussed earlier amniocentesis is a second trimester technique usually carried out between 16-18 weeks into the pregnancy. If cell culture is then carried out there will be a further delay of 2–4 weeks before results become available.

SAQ 5.1e

What do you think are the disadvantages and problems of this time sequence. As well as various scientific and medical considerations, put yourself in the place of the mother and try and imagine any problems of a more human kind that might arise.

SAQ 5.1e

To minimise the problems this time sequence can generate, the analysis of specimens from the chorionic plate is being increasingly used. The chorion is a multilayered sac outside the amnion which is involved in the absorption of nutrients before the placenta is well developed, and the chemical and genetic composition of the tissues does seem to be fetal rather than maternal.

Entry to the abdomen is usually via the cervix and samples are obtained either by a blind aspiration, a combination of washing and aspiration or most successfully by fetoscopy and/or ultrasound-guided needles or forceps. With this latter technique the success rate in obtaining fetal tissue is nearly 100%. The physical hazard to the mother is thought to be negligible, the increase in fetal spontaneous abortion rate is probably $<5\%$ and patient acceptability is considerably better than for standard amniocentesis techniques. While the samples obtained are small, the availability of miniaturised analytical techniques and particularly the development of refined genetic analyses will make the technique of great significance in the next few years.

5.2. APPLICATIONS OF THE FETAL BIOPSY TECHNIQUES

Perhaps three main areas of application have evolved:

(*a*) the study of gross anatomical defects;

(*b*) investigations of various diseases and physiological conditions affecting the fetus;

(*c*) investigations of inherited diseases and as an aspect of this, the determination of fetal sex.

(*a*) The technique of amniocentesis with fetoscopy has been of great value in anatomical studies since although ultrasound alone will identify some defects (eg the neural tube defects of anencephaly and spina bifida, which can be confirmed by studies on AFP concentration in the amniotic fluid), many defects are much more, or even only, apparent on direct visualisation. However a problem with fetoscopy in this regard is that the amniotic fluid tends to cloud after about 20 weeks giving a relatively narrow time-window for the procedure.

(*b*) An important application of these amniocentesis techniques is the determination of fetal maturity, particularly with regard to lung development. This is valuable as a warning of the possibility of the so-called respiratory distress syndrome of the newborn, which is one of the commonest causes of death in pre-term infants in the UK. Lung maturity is reflected in certain lipid changes in the amniotic fluid and fortunately these changes are relatively easily measured provided uncontaminated and properly stored samples are available. Another application is the prediction of cases of rhesus haemolytic disease of the new-born which has generally involved assays of bilirubin levels in the amniotic fluid, but investigations of more specific proteins (human placental lactogen, diamine oxidase and selected antibodies) might become more significant in the future. The seriousness of this condition was one of the main incentives for developing these techniques in the first place, and they have reduced the incidence of perinatal death due to it, to $<1\%$ of the total. The availability of fetal tissue and blood samples allows the investigation of fetal antibodies both for the diagnosis of immunodeficiency conditions, but also for fetal infections by Rubella and Cytomegalovirus which are indicated by changes in type M immunoglobulins.

(*c*) In a review in 1983, about 70 inherited diseases were said to be identifiable in the fetus although many of the results needed independent confirmation. In some cases direct chemical analysis of the fluid is possible, eg for complex polysaccharides produced in the mucopolysaccharidoses, and for certain enzyme deficiencies in the lysosomal storage diseases. In most cases cell components have to be studied either with or without preliminary culture. In a survey done in 1983 at King's College Hospital, London, out of 614 amniocenteses nearly half were for the investigation of possible defects in haemoglobin structure (haemoglobinopathies) many of which are hereditary, and another 130 were for haemophilia which is well-known as an inherited condition. It is undoubtedly the case that sophisticated genetic analyses involving gene probes etc will, in the near future, revolutionise these investigations since very small quantities of fetal tissue will be all that are required for the investigation. The fact that all nucleated cells contain all the genetic information of the individual will eliminate the cellular variability problem. Already, important conditions such as sickle cell anaemia and β-thalassaemia can be diagnosed by these techniques and probes are being developed for other important conditions such as Duchenne muscular dystrophy and haemophilia B.

Sex determination is important in the counselling of mothers with fetuses affected by sex-linked conditions, and identification of the Barr body in female cells and/or fluorescent labelling of the Y chromosome in males is possible. The reliability of these techniques is quite low and many centres prefer to carry out a full chromosomal (karyotype) analysis since this is not only more reliable in sex determination but will also identify many serious chromosomal aberrations, eg Down's and Turner's Syndromes. Lymphocytes obtained from fetal blood samples can be used for these studies and provide results $<$ 3 days after sampling.

Summary

Several different types of specimen can be removed from a pregnant woman in order to carry out investigations on her fetus. The nature of these specimens, the often intricate procedures required to obtain them, problems with the sampling procedures and the samples themselves are discussed and an indication of the main types of investigation given.

Objectives

You should now be able to:

- briefly describe the anatomy and physiology of a fetus and its environment in the uterus and hence state the principal types of specimen that can be obtained as an aid to its investigation;

- describe the techniques used to obtain samples of these materials and the problems involved in so doing;

- outline some of the main types of investigation carried out on these samples.

Revision Questions

The following set of questions are purely for revision purposes. They are based on material which has appeared in the text and most of them can be answered by simple statements.

The answers are given, without further explanations, on page 85.

1. Name the principal controlling influence on the movement of water across cell membranes.

2. What name is given to the equation which relates the cell membrane potential to intracellular and extracellular electrolyte concentrations?

3. Explain why the cell membrane potential cannot be determined directly given the intracellular and extracellular sodium ion concentrations.

4. In the blood capillaries some protein is lost from the plasma to the interstitial fluid. Where is it thought that most of this protein is returned to the blood?

5. State why the red blood cells have limited respiratory abilities compared with most other cells in the body.

6. What change in the appearance of the plasma is likely to be seen after a person has eaten a meal containing fats?

7. What reason was put forward to explain why vegetarians tend to excrete certain drugs and metabolites (eg urobilinogen) at a different rate from persons on a mixed diet?

8. Name the two reasons which were proposed to account for the differences in concentration of plasma constituents between males and females during the period from adolescence to middle age?

9. What three contributing factors could explain the seasonal changes which have been reported in the concentrations of blood constituents?

10. What effect has stress has on a individual's plasma cortisol levels?

11. Predict a possible consequence of acute anxiety on the blood pH.

12. Does the renal clearance of a substance increase, decrease or show no change with the size of an individual?

13. Would the concentration of proteins in the plasma increase or decrease after a person changes from a supine to an upright position?

14. What is the 'alkaline tide' which occurs after a meal is eaten?

15. Does capillary blood most closely resemble arterial or venous blood in its composition?

16. State the conditions under which there may be significant differences between the glucose concentrations of capillary and venous blood specimens.

17. List the practical disadvantages associated with the use of capillary blood compared with venous blood.

18. What effects does the use of a tourniquet have on the composition of venous blood?

19. Explain why plasma collected from blood which has been lightly centrifuged may show falsely high enzyme activities.

20. What are the advantages of using whole blood for assays in preference to plasma or serum?

21. Name the quantitatively most important protein which is present in plasma but not in serum.

22. What property do the sodium salts of EDTA, citric acid and oxalic acid have in common as anticoagulants?

23. What change is likely to occur in the pH of venous blood during storage?

24. Identical lactate dehydrogenase (LDH) activities were recorded from two different patients when their plasma specimens were analysed on the day of collection. The plasma samples were stored under identical conditions, and when they were analysed again on the following day, one sample showed no loss of activity while the other gave a significantly lower result. Can you explain these observations?

25. What are the three most relevant routes for the transmission of infection via blood specimens?

26. List four tissues and three fluids that are commonly sampled for diagnostic purposes.

27. Give one disease or condition which can be investigated using samples of (*i*) CSF, (*ii*) synovial fluid, (*iii*) gut tissue.

28. Briefly describe the main ways in which (*i*) urine, (*ii*) CSF and (*iii*) faecal samples from women, might be biologically contaminated.

29. State one type of change that can occur during the transport of (*i*) faeces, (*ii*) cells (*iii*) enzymes to the laboratory after sampling.

30. Organ biopsies are used for chemical (including enzymic) investigations. Name one other major type of investigation that is carried out on them and in fact is probably more important diagnostically.

31. Give two distinct examples of the use of fibre-optic devices to assist in obtaining different human samples.

32. In collecting saliva specimens why is it unsatisfactory for the patient simply to spit into a receptacle?

33. Lactose intolerance is a very common condition in which the absence of lactase results in a failure to metabolise lactose (the 'milk sugar'), giving the individual an unpleasant but not fatal diarrhoea. How might you use the biopsy technique to confirm this condition?

34. What is meant by the term 'circadian rhythms' and how do they affect urine investigations?

35. List the problems involved with collecting 24 hour urine samples.

36. Name one common urine preservative and state one problem involved with its use.

37. A sample of urine has been placed in a clear glass bottle on a window-sill for a few days. State three distinct types of change that might develop in this urine that might affect its value in diagnosis.

38. When urine is collected for cytological investigations, why is it desirable to wipe the external genitalia before the urine is excreted, and also to avoid the first 100 cm^3 or so of the sample?

39. In what ways does the fact that the lower part of the gut is contaminated with micro-organisms affect the value of faecal investigations in diagnosis?

40. Give two reasons why a single faecal specimen is of limited value for diagnostic investigations.

41. In the collection of faeces for occult blood investigations, why is it necessary for the patient to abstain from gut irritants such as aspirin?

42. What do the initials CSF stand for?

43. Give one advantage and one disadvantage of the procedure for sampling the CSF in which a sample is removed from the lumbar region of the vertebral column.

44. Describe the basic process of amniocentesis.

45. List the types of reported deleterious effect of amniocentesis on the fetus.

46. (*a*) Why is it necessary to wait for at least fourteen weeks before carrying out amniocentesis?

    (*b*) Bearing (*a*) in mind what are the problems with there being a legal deadline of twenty weeks for a therapeutic abortion?

47. List the advantages of fetoscopy.

48. What are the principal advantages of chorionic villus sampling?

49. State how fetal red blood cells can be distinguished from maternal ones.

50. Tay–Sach's disease can be investigated by measurements of the enzyme hexosaminidase A, which is significantly reduced in affected individuals. In prenatal studies the results are frequently expressed as a ratio of hexosaminidase A/A + B activities. Why is this?

Answers to the Revision Questions

1. Osmotic pressure.

2. The Nernst equation.

3. A large electrochemical gradient is maintained across the cell membrane by an active transport mechanism called the sodium pump. This gives far higher extracellular and lower intracellular sodium ion concentrations than are predicted by the Nernst equation.

4. Most of the proteins lost from capillary plasma to the interstitial fluid return to the blood via the lymphatic system.

5. Red blood cells lack mitochondria which contain the enzyme systems that allow complete oxidation of substances such as glucose.

6. The presence of chylomicrons will give the plasma a milky appearance for one to two hours after eating a meal containing fats.

7. The vegetarian diet tends to increase the urinary pH, which affects the excretion rate of pH-dependent substances such as urobilinogen.

8. The two principal reasons suggested were differences in hormone levels and muscle mass between males and females.

9. The three contributory factors suggested were

   (*a*) dietary differences, eg availability of fresh fruit and vegetables;

   (*b*) the calorific content of the diet which tends to be higher in winter;

   (*c*) differences in exposure to uv radiation relating to ambient temperatures and an individual's exposure to sunlight.

10. Plasma cortisol concentrations tend to increase as a result of stress.

11. At times of acute anxiety a person is liable to hyperventilate giving increased loss of carbon dioxide and a subsequent increase of the blood pH.

12. Renal clearance increases with body size.

13. Plasma proteins may increase in concentration by 10% to 20% as a result of changing to an upright posture.

14. The increase in the pH of venous blood leaving the stomach which is the result of gastric acid secretion.

15. The composition of capillary blood most closely resembles that of arterial blood.

16. The concentration of glucose in capillary blood may be significantly greater than that of venous blood after a meal containing carbohydrates has been taken. Venous and capillary bloods show similar glucose concentrations under fasting conditions.

17. There is a greater tendency for haemolysis and contamination with tissue debris to occur in capillary blood compared with venous blood. Capillary blood is generally available in smaller volumes than venous blood.

18. Use of a tourniquet tends to cause haemoconcentration of blood in the capillaries accompanied by increases in the concentrations of proteins and protein-associated substances in the plasma.

19. The plasma is likely to contain platelets with their complement of 'cellular' enzymes.

20. There are many disadvantages in using whole blood rather than plasma or serum but the former does provide a greater volume for analysis. It is also both more economic and presents a lower risk of infection to laboratory staff by virtue the fact that the separation stage is avoided.

21. Fibrinogen.

22. All three act by effectively removing calcium ions from the blood.

23. There is a tendency for the pH of venous blood to rise during storage as carbon dioxide is lost.

24. The most probable explanation is that the two specimens contained different relative amounts of the LDH isoenzymes, but the same total activity. Some isoenzymes degrade more rapidly than others giving the observed difference on the second day.

25. (*a*) percutaneous transmission by needle or other sharp object;

    (*b*) percutaneous transmission by contamination of cuts;

    (*c*) contamination of the mucosal surfaces of the mouth, nose or eyes.

26. (*i*) Tissues: liver, gut, lung, bone marrow, skin, fetal specimens.

    (*ii*) Fluids: blood, urine, faeces, saliva, CSF, lymph, pleural and synovial fluids.

27. (*i*) Cerebral infarction, inflammation, tissue degeneration and general trauma.

    (*ii*) Inflammation (including rheumatism, arthritis etc) and gout.

    (*iii*) Various malabsorption syndromes.

28. (*i*) Blood in menstruating women, vaginal secretions, tissue debris from the urinary tract, and virtually anything from the collection container.

    (*ii*) Blood and tissue from the body mass between the skin and the CSF.

    (*iii*) Urine.

29. (*i*) Bacterial metabolism of components, spontaneous oxidation of some constituents.

    (*ii*) A wide range of changes due to decrease in cell energy, breakdown of macromolecules, and degeneration of structures especially membranes.

    (*iii*) Denaturation, or inhibition due to the effect of preservatives or contamination from the container.

30. Histological and/or cytological studies.

31. Accurate location of sampling devices in the gut, and visual examination and again accurate location of sampling devices when obtaining fetal specimens.

32. The sample will be contaminated with fluids and cellular material from the mouth and upper part of the respiratory tract.

33. Tissue samples from the gut could be obtained and the presence or absence of the lactase enzyme demonstrated by microscopical examination following the reaction of tissue sections with a colour producing, enzyme specific stain (an enzyme histochemistry technique).

34. The influence of time on a factor; commonly (but not always) the factor varying in concentration across a 24 h period – hence 'circadian'. Many urine components show such variations.

35. Many of the problems actually encountered are not a consequence of the 24 hour period itself but tend to arise out of human nature and the lack of close supervision of the collection when done at home or even in a hospital ward:

    — contaminated containers might be used;

    — specimens might be contaminated with other biological materials;

— inappropriate or low concentration preservatives might be used;

— poor storage and transport conditions might exist;

— specimen might be lost;

— the time period for collection might be incorrect.

36. (*i*) Strong mineral acids eg HCl, or organic compounds such as thymol or formalin.

    (*ii*) In addition to the general fact that all preservatives interfere with some (although usually different) assay systems, the former is obviously potentially dangerous and some of the latter have poor miscibility with urine.

37. (*i*) microbial growth and metabolism of components;

    (*ii*) precipitation of urates;

    (*iii*) oxidation, eg of bilirubin, bilinogens and vitamin C;

    (*iv*) light activated degradation, eg of bilirubin, or various polymerisation reactions which result in melanin production.

38. To minimise contamination with cellular debris from the tissue surfaces over which the urine must pass before it can be collected.

39. The bacteria metabolise some of the faecal components in a poorly reproducible way.

40. Extensive variation in composition of a single faecal deposit occurs both between individuals and for a single individual. This is because of variations in general gut metabolism, extent of dehydration, percentage delivered, extent of microbial action and transit time, together with a number of other factors.

41. Slight leakage of blood due to capillary haemorrhage, can result from taking aspirin and similar materials and this makes diagnosis of other causes of leakage of blood into the gut less reliable.

42. Cerebrospinal fluid.

43. The spinal cord is not present and hence cannot be damaged, but the composition of the CSF is slightly different from that in the region of the brain.

44. The description ought to contain reference to:

    – the nature and location of the amniotic fluid and its containing membranes;

    – localisation of the optimum site, depth and route of needle insertion by ultrasound and/or fetoscopy;

    – the need for local anaesthesia only;

    – the various types of specimens that can be obtained and the devices used to obtain them.

45. – spontaneous abortion;

    – physical damage;

    – a possible increase in some respiratory and musculo-skeletal disorders seen in live-birth babies;

    – low birth-weight.

46. (*a*) To allow sufficient fluid to develop for a sample to be removed with minimum potential harm to the fetus.

    (*b*) Relatively little time is available for tests to be carried out, particularly those involving cell cultures, before the legal deadline for therapeutic abortion. The chance to repeat those tests that fail or are inconclusive is usually slight.

47. By allowing accurate location of the sampling devices, specimens of particular fetal materials can be obtained with minimum damage to the fetus and minimum contamination by maternal tissues and fluids. For example samples from the chorionic plate, placenta, fetal blood system or organs such as skin or liver can be obtained.

48. Samples can be obtained much earlier in the pregnancy thus minimising potential harm to the fetus and distress to the mother, and allowing longer for tests to be carried out before the legal deadline for therapeutic abortion.

49. Fetal blood cells are larger than maternal ones, and also have a different composition which allows them to be distinguished by a variety of differential precipitation reactions, some of which use specific antibodies, or by density gradient centrifugation.

50. The expression of results as ratios significantly reduces the need for complete, or at least highly reproducible, extractions of enzymes from samples, which is the case if a single enzyme activity is to be used as a diagnostic indicator. A poor extraction on a particular occasion would give an apparently low value if a single measurement is used and this could be regarded as suggestive of a clinical condition. In many situations the poor extraction would affect both components of a ratio similarly and thus a minimal or zero distortion of the ratio should occur.

Self Assessment Questions and Responses

SAQ 1.1a

To check your knowledge of osmotic pressure, and the features associated with it, write brief answers to the following questions.

(*i*) What is an ideal solution?

(*ii*) What does the osmotic pressure of a solution depend upon?

(*iii*) Osmotic pressure is one of the four colligative properties of a solution. Name the other three.

Response

(*i*) An ideal solute (in this context) is one that neither dissociates into two or more components, nor associates to reduce the number of particles in solution. Glucose at low concentrations could be described as an ideal solute unlike sodium chloride which dissociates in solution to give sodium and chloride ions.

(*ii*) Osmotic pressure depends upon the number of solute particles per unit volume of solvent and is independent of their chemical composition.

(*iii*) The other three colligative properties of a solution are its freezing point, boiling point and vapour pressure. The word 'colligative' indicates that the four properties are bound together and therefore a change in any one automatically means a change in all four. A colligative property is one that depends on the number of molecules or particles in the system rather than on their nature.

****

SAQ 1.1b

The membrane potential of a red blood cell is 0.008 V and the plasma chloride concentration is 100 mmol dm^{-3}. Using the simplified version of the Nernst equation (1.1c), calculate the predicted intracellular chloride concentration for the red blood cell.

Response

From Eq. 1.1c the following relationship applies:

$$E = 0.061 \log (c_i^- / c_e^-)$$

where c_i^- is the intracellular concentration of chloride in the red blood cell in mmol dm^{-3}.

$$0.008 = 0.061 \log c_i^-/100$$

$$\log c_i^- = \frac{0.008}{0.061} + 2$$

$$c_i^- = 135.2 \text{ mmol dm}^{-3}$$

****

SAQ 1.1c

Can you name any situations which are liable to result in depletion of ATP levels within the cell?

Response

Lack of substrate (eg glucose), anoxia (lack of oxygen) or the presence of metabolic poisons are all factors which are liable to deplete the supply of ATP.

Each of these factors may affect a blood specimen since the normal processes for supplying nutrients and removing waste products are absent. It is also possible that preservatives which have been added could interfere with the metabolic pathways that generate ATP.

****

SAQ 2.1a

Given the following information about the major protein fractions found in plasma, explain briefly which of them is likely to play the most important part in the control of blood volume.

Remember that the control of blood volume by the plasma proteins is due to their containment within the vascular fluid where they account for the colloid osmotic pressure. ⟶

SAQ 2.1a (cont.)

Protein	Plasma levels ($g\ dm^3$)	Relative molecular mass
albumin	40.0	66 300
α_1-globulins	4.5	40 000–60 000
α_2-globulins	6.5	100 000–400 000
β-globulins	8.5	110 000–120 000
γ-globulins	11.0	15 000–400 000
fibrinogen	3.0	340 000

Response

The answer is albumin.

Osmotic pressure depends upon the number of solute particles per unit volume of solvent. The table shows that albumin has the highest concentration in $g\ dm^{-3}$ (greater than the total globulin concentration) and has the lowest relative molecular mass, save for the α_1-globulins. The albumin fraction therefore contains a far greater number of individual solute molecules per unit volume.

Please note that the nomenclature used (α_1, α_2, β, γ) is based upon the globulin fractions which can be differentiated by low voltage electrophoresis at pH 8.6. Each of the globulins represents a mixture of different protein species in contrast to the albumin and fibrinogen fractions which are relatively pure.

****

SAQ 2.2a

Reports indicate that serum cholesterol concentrations tend to be higher in the winter than in the summer. Can you list three possible explanations for this finding? The formula given below for cholesterol may give you a clue for one of them.

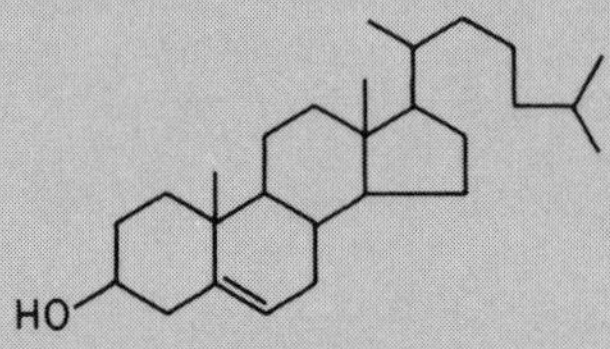

HO

Cholesterol

Response

(*i*) Increased exposure to ultraviolet irradiation during the summer may cause a reduction of cholesterol in the body through oxidation. Note the similarity between cholesterol and the vitamin D precursor (7-dehydrocholesterol).

(*ii*) The calorific value of food taken is often greater in the winter.

(*iii*) People often undertake more recreational physical activities during the summer months. Exercise is recognised as a factor which tends to lower serum cholesterol levels.

Although the first explanation is valid, it is likely that (*ii*) and (*iii*) account for most of the differences that have been reported.

****

SAQ 2.2b

It was stated (2.1) that some of the plasma calcium is protein-bound and the rest is ionised or complexed with small substances such as carbonate or citrate.

A volunteer was asked to rest on a bed for three hours and a blood specimen was withdrawn at the end of that period while he remained in a recumbant position. Tests on the serum of that specimen showed a total protein concentration of 66.3 g dm^{-3} and a total calcium concentration of 2.30 mmol dm^{-3} of which 40% was protein-bound. The volunteer was asked to walk about for 20 minutes and a second blood specimen was then collected.

Make an estimate of the total serum calcium concentration of the second specimen which had a total serum protein value of 76.5 g dm^{-3}.

Note any assumptions that are necessary when making your estimate of the new calcium concentration.

Response

40% of the calcium in the first serum specimen represents $2.30 \times 40/100$ or 0.92 mmol dm^{-3} which leaves 2.30–0.92 or 1.38 mmol dm^{-3} as the non-protein-bound fraction.

The change from the recumbent to the upright posture increases the concentration of protein-bound substances by a factor of 76.5/66.3. The protein-bound calcium concentration of the volunteer will therefore become $0.92 \times 76.5/66.3$ or 1.06 mmol dm^{-3}.

Assuming that the concentration of non-protein-bound calcium is unaltered, the serum total calcium concentration of the second specimen is 1.06 + 1.38 or 2.44 mmol dm^{-3}.

Aside from assuming that the non-protein-bound calcium concentration is unaffected by the change in posture, we cannot be sure that all the plasma proteins remain in the vascular spaces. Indeed it has been found that protein does pass into the interstitial fluid and that most of that protein is albumin. Albumin happens to be the protein which binds most of the protein-bound fraction of the serum calcium.

****

SAQ 3.1a

Based on the results of Eriksson and his colleagues (Fig. 3.1a) suggest a set of recommendations relating to the collection of capillary and venous blood specimens when performing glucose tolerance tests.

Response

It might be reasonable to suggest the following:

- that either capillary or venous blood, but not both, be used during any one glucose tolerance test on a patient;
- that the laboratory provides examples of the values expected from both capillary and venous blood in glucose tolerance tests;
- that the physicians who will interpret the results be told whether it was capillary or venous blood which was used.

****

SAQ 3.1b

A tourniquet is applied to a volunteer's arm and a needle inserted in the median cubital vein. Exactly 5 cm^3 of blood is allowed to run at about 10 cm^3 min^{-1} into each of four bottles labelled 1, 2, 3 and 4 respectively. The blood samples are allowed to clot and albumin determined in the serum. Results as follows:

Specimen number	1	2	3	4
Albumin (g dm^{-3})	43	44	46	49

(*i*) Can you explain the increase of serum albumin in terms of the tourniquet effect?

(*ii*) Which of the samples is likely to be the most representative of the circulating blood?

(*iii*) In drawing up recommendations for venous blood collection are there any points which you would make relating to volumes of blood to be collected for specific analyses?

Remember that the tourniquet will increase intravascular pressure, and this in turn will increase filtration pressure across the capillary walls.

Response

(*i*) An increase in the intravascular pressure and the filtration pressure across the capillary walls causes movement of water and the low molecular mass substances into the interstitial fluid. This results in an increase of the plasma concentrations of proteins and any protein-bound substances.

(*ii*) The blood which is already in the vein near to the tourniquet (sample number 1) is the most representative of circulating blood. Later blood is more likely to have originated from the small veins and the capillaries where exchange with the interstitial fluid can occur.

(*iii*) It would be reasonable to recommend a policy of collecting minimal sample volumes. When larger volumes are needed for several analyses and different sample containers are used, the first-drawn blood should be reserved for the most critical of those analyses, eg calcium.

A standardised method of blood collection should be adopted which specifically states how tourniquets should be applied and for how long.

****

SAQ 3.2a

The following set of plasma values were obtained for an apparently normal person. There are grounds for suspecting that the blood specimen was collected into an inappropriate anticoagulant.

Try to identify which anticoagulant(s) described previously could have been used. ⟶

SAQ 3.2a (cont.)

Analyte	Units	Result	Reference value
Potassium	mmol dm^{-3}	6.3	3.7–5.2
Sodium	"	139	135–146
Calcium	"	1.9	2.3–2.7
Bicarbonate	"	24	22–26
Chloride	"	103	98–108
Phosphate	"	1.12	0.7–1.4
Urea	"	4.9	3.3–7.5
Total protein	g dm^{-3}	72	68–79
Albumin	"	44	40–48

Response

A high potassium and low calcium suggests that the anticoagulant was a potassium salt which acted by binding calcium in some way. The dipotassium salt of EDTA or potassium oxalate could have produced the observed effects.

****

SAQ 3.4a

5 cm^3 of venous blood is placed in an anticoagulant- containing bottle which has a total volume of 15 cm^3. This leaves a considerable air space above the specimen.

Try to predict whether the concentrations of the following substances are likely to increase, decrease or remain constant in both the red cells and the plasma as a result of mixing the specimen with air. The behaviour of chloride ions is not easily identified. Try to remember that it is not just the balance of each individual species between cells and plasma which is important, but also the balance between charged species.

Substance	Cells	Plasma
oxygen (O_2)		
carbon dioxide (CO_2)		
bicarbonate (HCO_3^-)		
chloride (Cl^-)		

The following information may help your decisions:

(*i*) the red cell membrane is permeable to CO_2, O_2, HCO_3^-, Cl^-, and H^+;

(*ii*) loss of a substance from cells will be at least partially compensated by movement of that substance from the plasma and vice versa; ⟶

SAQ 3.4a (cont.)

(*iii*) although intra- and extra-cellular concentrations of individual substances may vary, the balance between cations and anions on either side of a cell membrane is usually maintained;

(*iv*) red cells contain the enzyme carbonic anhydrase which catalyses the reversible reaction between hydrogen and bicarbonate ions to form carbonic acid leading to carbon dioxide and water:

$$H^+ + HCO_3^- \rightleftharpoons H_2CO_3 \rightleftharpoons CO_2 + H_2O$$

Remember that enzymes increase the rate of reaction towards equilibrium states;

(*v*) when oxygen combines with haemoglobin a proton (H^+) is displaced. The reverse occurs at low pO_2 values when a proton is taken up by haemoglobin as oxygen is released.

(*vi*) compared with room air venous blood is relatively deficient in oxygen and rich in carbon dioxide.

Response

There is a net gain of oxygen which diffuses from air to plasma and from the plasma into the red blood cells. In the red cells it combines with haemoglobin and in the process displaces a proton.

Carbon dioxide diffuses from plasma to the air. This loss from the plasma is partially compensated by movement of carbon dioxide out of the red cells. The result is a loss of carbon dioxide from the blood.

Reduction of intracellular carbon dioxide and an increase in the number of hydrogen ions drives the carbonic anhydrase catalysed reaction to the right. This reduces the intracellular concentration of bicarbonate and results in movement of plasma bicarbonate into the cells to maintain intra-extracellular balance.

All the changes described so far, are illustrated in the diagram, follow the diffusion of carbon dioxide and oxygen out of and into the specimen respectively. How then is the movement of chloride involved?

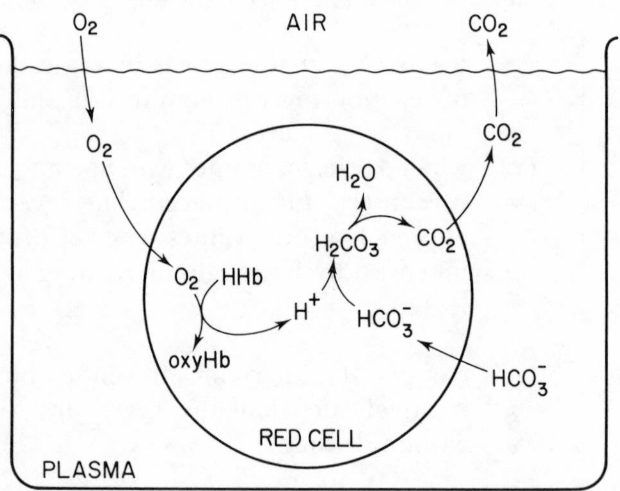

The diagram shows loss of an anion (HCO_3^-) from the plasma to the cells which tends to give an overall ionic imbalance between the two (remember that the proton is derived from haemoglobin). Partially to compensate for this imbalance there is a net movement of chloride ions from cells to plasma. This is the phenomenon called the 'chloride shift' or 'Hamburger reaction'.

In practice the variations in chloride concentrations are small whereas significant changes are seen with the other parameters.

****

SAQ 3.4b

Respiration is often summarised as the oxidation of glucose to carbon dioxide and water:

$$C_6H_{12}O_6 + 6\,O_2 \rightarrow 6\,CO_2 + 6\,H_2O + \text{energy}$$

In stored blood however, the main product of glucose catabolism is lactic acid:

$$C_6H_{12}O_6 \rightarrow 2\,C_3H_6O_3$$

Which of the following statements offers the best explanation for the difference between glucose breakdown in the blood compared with that in other tissues?

(*i*) Red cells metabolise glucose by a different pathway from that used by other cells of the body.

(*ii*) Glucose catabolism is a two stage process. The first stage takes place in the blood, and the second stage in the tissues where the lactic acid is broken down to carbon dioxide and water.

(*iii*) Red cells lack mitochondria, hence they are unable to oxidise glucose completely.

(*iv*) Complete oxidation of glucose to carbon dioxide and water requires the presence of oxygen. In stored blood the supply of oxygen is limited since the cells settle at the bottom of the bottle and are separated from the air by a layer of plasma.

Response

Statement (*ii*) is the most satisfactory explanation of the four. Most cells catabolise glucose in the presence of oxygen via pyruvate to carbon dioxide and water giving a relatively high energy yield in terms of ATP and related compounds. The glycolytic pathway leading to pyruvate takes place in the cytosol of the cell. Krebs cycle and the respiratory chain then complete the oxidation to carbon dioxide and water within the mitochondria.

Red cells lack mitochondria and must therefore rely on the glycolytic pathway and the small net gain of ATP which it supplies. Conversion of pyruvate to lactate replenishes the supply of the oxidised form of nicotinamide adenine dinucleotide (NAD^+) which is necessary for one of the stages in the glycolytic sequence of reactions:

$$C_3H_4O_3 + NADH + H^+ \rightleftharpoons C_3H_6O_3 + NAD^+$$

The conversion of glucose to lactate with the net gain of 2 molecules of ATP is the means by which cells, which do possess mitochondria, continue to respire under anaerobic conditions. Please remember that glycolysis is not the only means by which glucose is metabolised by the red cell. Another important pathway in the red cell is the hexose monophosphate shunt which does actually produce carbon dioxide but does not give a direct energy gain in terms of ATP etc.

Sodium fluoride is commonly used at a final blood concentration of 2 mg cm^{-3} to prevent glycolysis along with an anticoagulant such as potassium oxalate or EDTA. Some laboratories use sodium iodoactetate in preference to fluoride on the grounds that the latter tends to be inhibitory to a wide range of enzymes, eg urease which is used in a blood urea assay method.

****

SAQ 3.4c

Absorption spectra for bilirubin and haemoglobin in phosphate buffer solutions are shown in Fig. 3.4b and 3.4c respectively.

Can you suggest:

(*i*) suitable wavelength for reading the absorbance value of bilirubin, and

(*ii*) a simple method of correcting the bilirubin absorbance for interference by any haemoglobin which might be present in a serum sample.

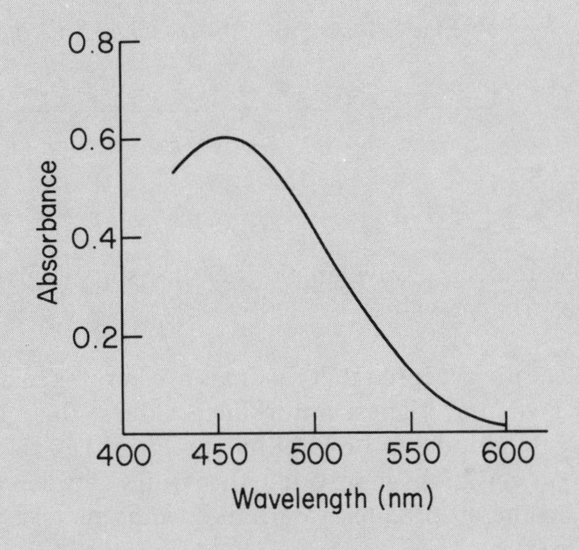

Fig. 3.4b. *Absorption spectrum for bilirubin*

→

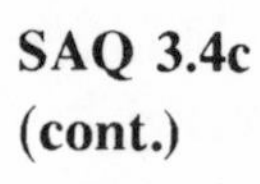

SAQ 3.4c (cont.)

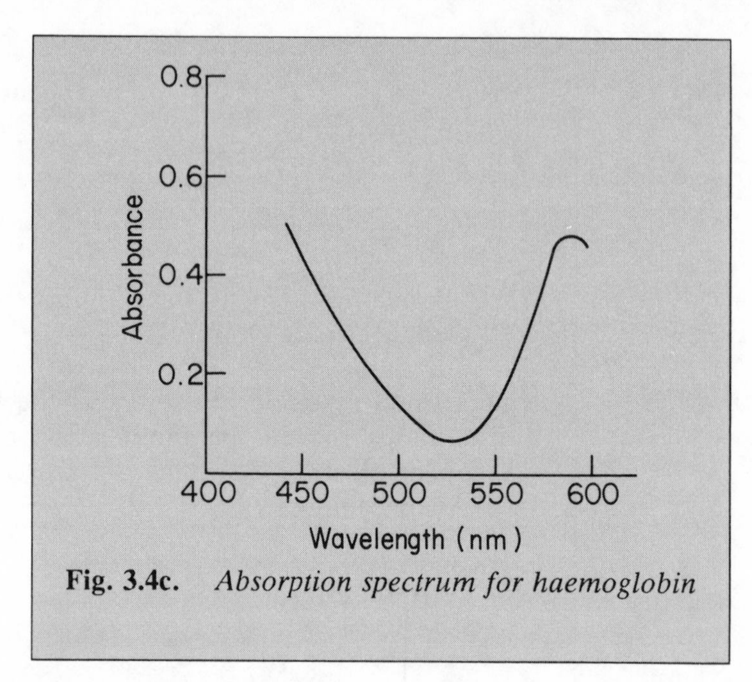

Fig. 3.4c. *Absorption spectrum for haemoglobin*

Response

(*i*) 455 nm is the wavelength of choice since it gives the maximum absorbance value.

It is generally accepted that readings are best taken at the wavelength which gives the highest absorbance values; this gives the greatest sensitivity. Good reproducibility (precision) is also obtained since any small variations in wavelength settings between assays have little effect on the absorbance values due to the plateau shape at the top of the peak.

(*ii*) At 455 nm haemoglobin also absorbs strongly and would interfere with bilirubin readings taken at this same wavelength. However, a similar absorbance is given by haemoglobin at 550 nm where the absorbance due to bilirubin is very low. By taking the difference between absorbances at the two wavelengths (A_{455}–A_{550}) an approximation of the absorbance due to bilirubin alone at 455 nm is obtained.

It is important to remember that absorption spectra may vary slightly when measured in different spectrophotometers and you should always be prepared to establish wavelength optima on your own instrument rather than simply accept published results.

****

SAQ 4.0a

What do you think are the more significant scientific problems with obtaining biopsy specimens?

Response

Scientific problems include the need to take a truly representative sample, avoidance of the need for repeat sampling, and ensuring adequate sample preservation. This is particularly important for histological work which unfortunately frequently has different and sometimes incompatible preservation requirements compared with those for biochemical studies. The actual investigation and evaluation is also troublesome for a number of reasons, eg only small sample volumes are usually available and a wide range of different types of investigation may be needed on these samples. Analytical reference data are frequently different for these samples and the determination of reference ranges for these sample types is made quite difficult by the relatively low number of normal sample specimens generally available. Lastly the very different biological nature of the specimens often means that methods evolved for the investigation of serum or urine are prone to problems when used for tissue or other fluid specimens.

****

SAQ 4.1a

There are three major types of jaundice:

(*i*) pre-hepatic, in which excessive red blood cell breakdown occurs and sufficient haemoglobin is produced to overcome the ability of the liver to remove it for breakdown and excretion in the bile;

(*ii*) hepatic, in which the liver cells are defective and cannot convert the haemoglobin properly or at an adequate rate into the derivatives excreted in the bile;

(*iii*) post-hepatic, in which the bile is unable to escape from the liver usually due to a blocked bile duct.

When other investigations indicate that the patient has pre-hepatic jaundice it is not really worthwhile carrying out a liver biopsy. Why do you think this is?

Response

In principle only in cases of hepatic jaundice are the liver cells metabolically defective, which would be confirmed by suitable enzyme investigations on biopsy specimens.

****

SAQ 4.1b

Bile is a relatively toxic material and hence it is sometimes useful to investigate a liver biopsy when studying a serious or prolonged case of post-hepatic jaundice. What is the justification for this?

Response

Unfortunately, prolonged high levels of bilirubin from bile in the blood can damage some cells (including those of the brain of infants who have neonatal jaundice), and in this case the cells of the liver might ultimately become damaged. Such a possibility could warrant the analysis of a liver biopsy notwithstanding the problems involved in obtaining it.

****

SAQ 4.2a

Single faecal samples have a number of disadvantages from an analytical point of view. Consider what they might be.

Response

Whereas simple measurements for the presence of materials, eg urobilinogen, can be done on any faecal specimen, for the majority of investigations this is inadequate, partly because a number of important factors in alimentary physiology are very variable, even in normal individuals. Among these are the transit time from duodenum to rectum, the extent of dehydration of the faeces, the percentage of inert fibre and gas, and the timing and completion of rectal emptying. Studies on single individuals have shown that consecutive 24 h collections may give faecal fat measurements with as much as 100% variation in value.

****

SAQ 4.2b

When collecting these longer term specimens a problem exists in correlating faecal deliveries with the chosen time period for analysis. Orally taken dyes are frequently used to identify the beginning and end of the chosen collection period. How can this be done?

Response

For several decades, dyes have been used as an aid in marking the transit of material through the gut, by identifying the beginning and end of the timing period. Thus a material such as carmine red, gentian violet or charcoal would be fed at a convenient starting point and a different dye fed after the required time delay (2, 3, 4 days etc). When collecting the specimens the deposit containing the first dye would be included, whereas the deposit containing the second dye would be excluded and all intervening material pooled to form a total collection.

****

SAQ 4.2c

A faecal deposit of 200 g is homogenised in 1 dm^3 of water and produces 1250 cm^3 of homogenate. A 750 cm^3 volume of homogenate is investigated and has a Cr_2O_3 concentration of 1 mg cm^{-3}. Assuming the patient was supplied with 1.5 g of Cr_2O_3 as described above, what proportion of a days output of faeces does the 750 cm^3 represent?

Response

A Cr_2O_3 concentration of 1 mg cm^{-3} would give a total quantity in the aliquot of 750 mg, which is 50% of that supplied, and hence the analyst is dealing with 50% of a days output by the patient. To simplify matters no allowance has been made in these calculations for the effect of analytical inaccuracies, consistent bias etc.

****

SAQ 4.2d

Even if a material is excreted in urine evenly throughout the day there will be a variation in concentration due to differences in volume produced between day and night. Would an early morning or early evening specimen be the most dilute?

Response

Assuming only two collections were made in a single day then the concentration of an evenly secreted material will be lower in an early evening specimen (say 6–8pm) than in an early morning one (say 6–9am) due to this dilution effect.

****

SAQ 4.2e

Bearing in mind the procedure involved in collecting CSF, with what might the collected fluid be contaminated, and what effects might these contaminants have on the clinical value and reliability of the investigation?

Response

It is very important to minimise contamination of the CSF sample with blood and tissues from outside the CNS collected while the needle is inserted or withdrawn. This is important since the CSF might need to be examined for blood released due to physical damage or haemorrhage, and might also be investigated for changes in cell composition (particularly white blood cells) as this can be an important indicator of infection and inflammation.

****

SAQ 4.2f

Use your knowledge of general chemical analysis to suggest some of the different ways in which the increased level of sodium chloride in the sweat of individuals with cystic fibrosis might be shown, at least in a semi-quantitative way.

Response

Among the more common techniques are:

- Na measurement by a flame spectroscopy technique;
- Na^+ or Cl^- measurement by ion selective electrodes;
- Cl^- measurement by titration;
- NaCl measurement by osmometry.

****

SAQ 5.1a

An ultrasound scan of the mother's abdomen is extremely useful to the surgical team for a number of reasons. Bearing in mind what is actually being done to the mother, what do you think are the more straightforward of these benefits?

Response

Among other things it will improve their ability to:

- locate the best site for insertion of the needle;
- locate the best route for traverse of the needle through the mother's abdomen in order to minimise the chances of fetal or placental damage;
- identify the optimum depth for needle insertion;
- investigate fetal number, size, age and normality.

****

SAQ 5.1b

Why do you think this point concerning lung surfactant investigations is important, and can you imagine (in very simple terms) how the procedure might be modified to minimise it?

Response

The references cited elsewhere will give considerable detail of the reasons why fetal lung surfactant is investigated and the various methods available. One general approach is to study the effect its secretion into the amniotic fluid has on the lipid composition of the latter. Contamination of the fluid with other materials containing lipids will obviously affect its lipid composition and make investigations that much more difficult. To minimise one source of contamination the needle used to take the biopsy is guarded by a stilette which prevents the withdrawal of maternal material as the needle passes across the maternal abdomen. The extent of contamination of the fluid by fetal blood can be monitored by a measurement of the level of fetal haemoglobin (type F).

****

SAQ 5.1c

Cellular enzyme activities vary with a number of factors and you could consider what you think some of the factors contributing to this variation in cellular enzyme level might be. While you may be unfamiliar with the biological processes employed, just bear in mind that basically they involve placing the cells in a suitable growth medium, allowing them to grow, divide and multiply with periodic replenishment of the medium, until sufficient cells can be harvested for analysis.

Response

The article by Holton in Barson and Davis, (1981), gives a thorough review of this subject. The most significant are the effects of:

- the number of sub-culture stages employed;
- the phase of the cell cycle when the cells are studied;
- the culture conditions, including media composition, timing of media changes etc;
- cell type.

****

SAQ 5.1d

> There are perhaps two main applications of the fetoscopy technique; in very simple terms what do you think they are?

Response

- The identification of gross anatomical defects and fetal sex by direct inspection.
- An improvement in the safety of the collection of fetal blood or tissue samples. Early techniques for this collection involved repeated blind punctures of the placental plate but the technique had a fetal death rate of 7–10%. The combination of fetoscopy and ultrasound has reduced this level to < 2% above normal.

****

SAQ 5.1e

What do you think are the disadvantages and problems of this time sequence. As well as various scientific and medical considerations, put yourself in the place of the mother and try and imagine any problems of a more human kind that might arise.

Response

By 18 or 20 weeks into the pregnancy the mother will have become very well aware of the pregnancy particularly since the fetus may have begun its spontaneous movements. An earlier (ie first trimester) diagnosis would be useful in reducing the emotional upset involved in the investigation and possible termination. It would also allow a termination procedure that is technically easier and physically less traumatic for the mother. With these advantages in mind the technique of chorionic biopsy sampling is being evolved.

****

Units of Measurement

For historic reasons a number of different units of measurement have evolved to express quantity of the same thing. In the 1960s, many international scientific bodies recommended the standardisation of names and symbols and the adoption universally of a coherent set of units—the SI units (Système Internationale d'Unités)—based on the definition of five basic units: metre (m); kilogram (kg); second (s); ampere (A); mole (mol); and candela (cd).

The earlier literature references and some of the older text books, naturally use the older units. Even now many practicing scientists have not adopted the SI unit as their working unit. It is therefore necessary to know of the older units and be able to interconvert with SI units.

In this series of texts SI units are used as standard practice. However in areas of activity where their use has not become general practice, eg biologically based laboratories, the earlier defined units are used. This is explained in the study guide to each unit.

Table 1 shows some symbols and abbreviations commonly used in analytical chemistry; Table 2 shows some of the alternative methods for expressing the values of physical quantities and the relationship to the value in SI units.

More details and definition of other units may be found in the *Manual of Symbols and Terminology for Physicochemical Quantities and Units*, Whiffen, 1979, Pergamon Press.

Table 1 *Symbols and Abbreviations Commonly used in Analytical Chemistry*

Symbol	Meaning
Å	Angstrom
$A_r(X)$	relative atomic mass of X
A	ampere
E or U	energy
G	Gibbs free energy (function)
H	enthalpy
J	joule
K	kelvin (273.15 + t °C)
K	equilibrium constant (with subscripts p, c, therm etc.)
K_a, K_b	acid and base ionisation constants
$M_r(X)$	relative molecular mass of X
N	newton (SI unit of force)
P	total pressure
s	standard deviation
T	temperature/K
V	volume
V	volt (J A^{-1} s^{-1})
a, $a(A)$	activity, activity of A
c	concentration/ mol dm^{-3}
e	electron
g	gramme
i	current
s	second
t	temperature / °C
bp	boiling point
fp	freezing point
mp	melting point
≈	approximately equal to
<	less than
>	greater than
e, exp(x)	exponential of x
ln x	natural logarithm of x; ln x = 2.303 log x
log x	common logarithm of x to base 10

Table 2 *Alternative Methods of Expressing Various Physical Quantities*

1. **Mass (SI unit : kg)**

$$\text{g} = 10^{-3}\ \text{kg}$$
$$\text{mg} = 10^{-3}\ \text{g} = 10^{-6}\ \text{kg}$$
$$\mu\text{g} = 10^{-6}\ \text{g} = 10^{-9}\ \text{kg}$$

2. **Length (SI unit : m)**

$$\text{cm} = 10^{-2}\ \text{m}$$
$$\text{Å} = 10^{-10}\ \text{m}$$
$$\text{nm} = 10^{-9}\ \text{m} = 10\text{Å}$$
$$\text{pm} = 10^{-12}\ \text{m} = 10^{-2}\ \text{Å}$$

3. **Volume (SI unit : m^3)**

$$\text{l} = \text{dm}^3 = 10^{-3}\ \text{m}^3$$
$$\text{ml} = \text{cm}^3 = 10^{-6}\ \text{m}^3$$
$$\mu\text{l} = 10^{-3}\ \text{cm}^3$$

4. **Concentration (SI units : mol m^{-3})**

$$\text{M} = \text{mol l}^{-1} = \text{mol dm}^{-3} = 10^3\ \text{mol m}^{-3}$$
$$\text{mg l}^{-1} = \mu\text{g cm}^{-3} = \text{ppm} = 10^{-3}\ \text{g dm}^{-3}$$
$$\mu\text{g g}^{-1} = \text{ppm} = 10^{-6}\ \text{g g}^{-1}$$
$$\text{ng cm}^{-3} = 10^{-6}\ \text{g dm}^{-3}$$
$$\text{ng dm}^{-3} = \text{pg cm}^{-3}$$
$$\text{pg g}^{-1} = \text{ppb} = 10^{-12}\ \text{g g}^{-1}$$
$$\text{mg\%} = 10^{-2}\ \text{g dm}^{-3}$$
$$\mu\text{g\%} = 10^{-5}\ \text{g dm}^{-3}$$

5. **Pressure (SI unit : N m^{-2} = kg m^{-1} s^{-2})**

$$\text{Pa} = \text{Nm}^{-2}$$
$$\text{atmos} = 101\ 325\ \text{N m}^{-2}$$
$$\text{bar} = 10^5\ \text{N m}^{-2}$$
$$\text{torr} = \text{mmHg} = 133.322\ \text{N m}^{-2}$$

6. **Energy (SI unit : J = kg m^2 s^{-2})**

$$\text{cal} = 4.184\ \text{J}$$
$$\text{erg} = 10^{-7}\ \text{J}$$
$$\text{eV} = 1.602 \times 10^{-19}\ \text{J}$$

Table 3 *Prefixes for SI Units*

Fraction	Prefix	Symbol
10^{-1}	deci	d
10^{-2}	centi	c
10^{-3}	milli	m
10^{-6}	micro	μ
10^{-9}	nano	n
10^{-12}	pico	p
10^{-15}	femto	f
10^{-18}	atto	a

Multiple	Prefix	Symbol
10	deka	da
10^{2}	hecto	h
10^{3}	kilo	k
10^{6}	mega	M
10^{9}	giga	G
10^{12}	tera	T
10^{15}	peta	P
10^{18}	exa	E

Table 4 *Recommended Values of Physical Constants*

Physical constant	Symbol	Value
acceleration due to gravity	g	$9.81\ \mathrm{m\ s^{-2}}$
Avogadro constant	N_A	$6.022\ 05 \times 10^{23}\ \mathrm{mol^{-1}}$
Boltzmann constant	k	$1.380\ 66 \times 10^{-23}\ \mathrm{J\ K^{-1}}$
charge to mass ratio	e/m	$1.758\ 796 \times 10^{11}\ \mathrm{C\ kg^{-1}}$
electronic charge	e	$1.602\ 19 \times 10^{-19}\ \mathrm{C}$
Faraday constant	F	$9.648\ 46 \times 10^{4}\ \mathrm{C\ mol^{-1}}$
gas constant	R	$8.314\ \mathrm{J\ K^{-1}\ mol^{-1}}$
'ice-point' temperature	T_{ice}	273.150 K exactly
molar volume of ideal gas (stp)	V_m	$2.241\ 38 \times 10^{-2}\ \mathrm{m^3\ mol^{-1}}$
permittivity of a vacuum	ϵ_o	$8.854\ 188 \times 10^{-12}\ \mathrm{kg^{-1}\ m^{-3}\ s^4\ A^2\ (F\ m^{-1})}$
Planck constant	h	$6.626\ 2 \times 10^{-34}\ \mathrm{J\ s}$
standard atmosphere pressure	p	$101\ 325\ \mathrm{N\ m^{-2}}$ exactly
atomic mass unit	m_u	$1.660\ 566 \times 10^{-27}\ \mathrm{kg}$
speed of light in a vacuum	c	$2.997\ 925 \times 10^{8}\ \mathrm{m\ s^{-1}}$